STRAIGHT *A* STUDENT

explains

Non verbal reasoning

STRAIGHT A STUDENT
explains
Non verbal reasoning

N Robinson

authorHOUSE®

AuthorHouse™
1663 Liberty Drive
Bloomington, IN 47403
www.authorhouse.com
Phone: 1-800-839-8640

Published by AuthorHouse 07/09/2012

ISBN: 978-1-4772-1528-9 (sc)
ISBN: 978-1-4772-1529-6 (e)

CONTENTS

Introducing this Workbook 1

Section 1. Number Patterns and Positions 3

Section 2. Reflective Symmetry 8

Section 3. Rotational Symmetry 17

Section 4. Similarity and Congruency 25

Section 5. Mixed Non Verbal Reasoning 31

About The Author 45

Introducing this Workbook

What is non verbal reasoning?

This is the ability to understand and solve problems from pictorial representations of patterns and then go on to predict further images that follow the same patterns. These patterns may follow a numerical, reflective, rotational sequence or use similarity and congruency within the representations and then finally combinations of these relationships between the pictorial diagrams.

About this Workbook.

The booklet is in 5 sections. All sections are key to learning non verbal reasoning skills. Follow the instructions or examples in each section and complete practice worksheets and complete activities designed to improve learning. Answers to worksheets and skill practice questions are given at the back of this booklet. All worksheets and skill practice questions cover the syllabus for non verbal reasoning tests and their completion provides essential practice to encourage familiarity with entry tests.
They also offer essential tutoring for the required independent learning and thinking strategies for early secondary education and give a child a real boost in what is now a common ground requirement for their future educational needs and beyond.

Workbook contents.

Section 1. Numerical Patterns and Positions. This shows how patterns are built and diagrams form patterns with a number sequence as the basis or position of the shaded parts in a grid.

Section 2. Reflection Symmetry. This shows you basic symmetry skills and looks at shapes which are reflected in 1 or 2 lines of symmetry and also shapes which have a number of lines of symmetry. In this section there is a selection of worksheets which will coach non verbal questioning to solve problems using reflection as the key skill.

Section 3. Rotation Symmetry. This looks at the different types of rotation symmetry or order of rotational symmetry.

Section 4. Similarity and Congruence. This looks at shapes which are enlarged or reduced in size and their properties when: classifying, finding similar shapes or matching patterns of similar shapes.

Section 5. Mixed Non Verbal worksheets. Furthering the development of reasoning skills.

All sections coach the following thinking skills which are essential for success in non verbal reasoning techniques.

What are Thinking Skills?

Thinking skills are the elementary steps taken to break down a problem and use reasoning to acquire a solution or answer. The process may involve one or more of the skills mentioned below. This booklet promotes learning and problem solving methods as shown in the following spider diagram:

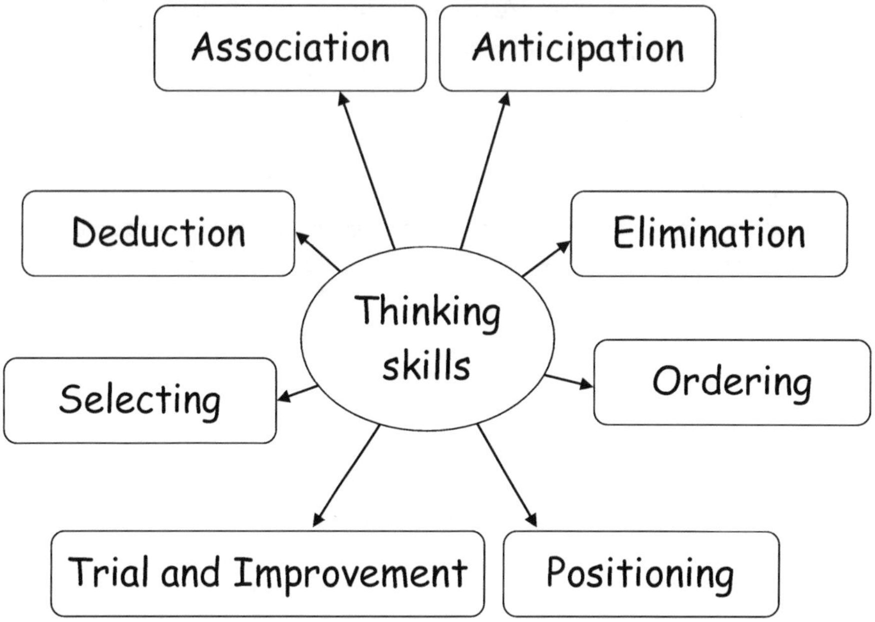

What do they all mean?

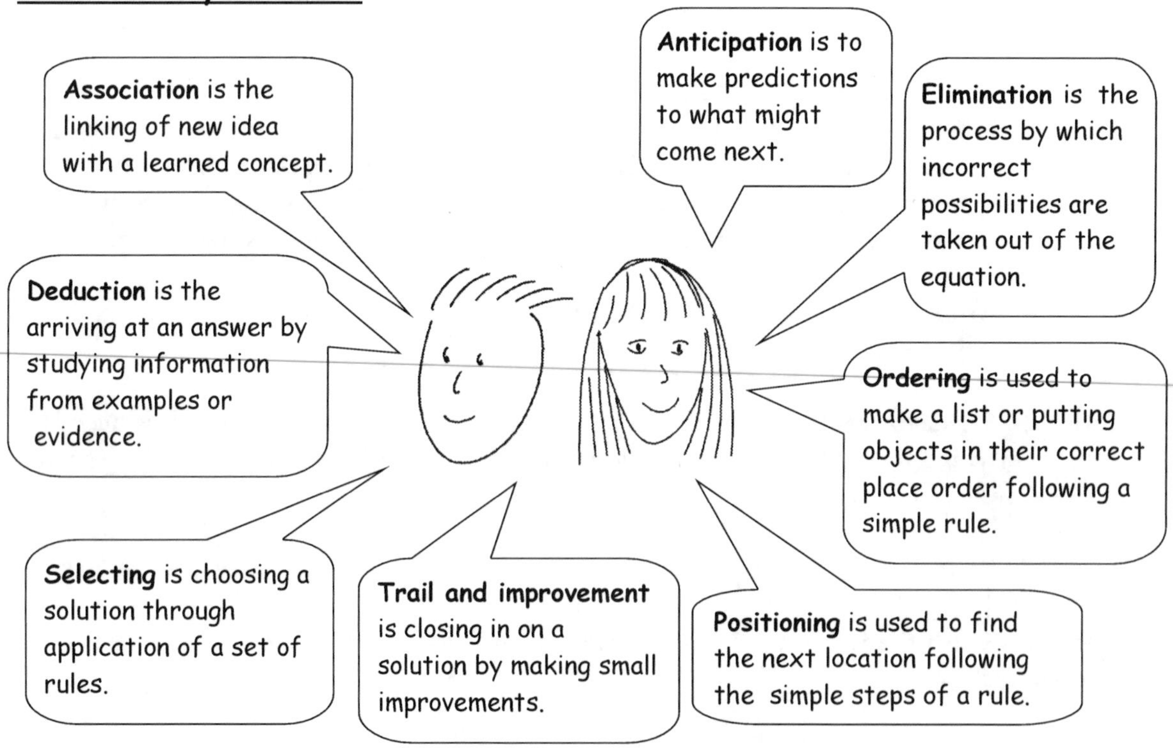

Association is the linking of new idea with a learned concept.
Anticipation is to make predictions to what might come next.
Elimination is the process by which incorrect possibilities are taken out of the equation.
Deduction is the arriving at an answer by studying information from examples or evidence.
Ordering is used to make a list or putting objects in their correct place order following a simple rule.
Selecting is choosing a solution through application of a set of rules.
Trail and improvement is closing in on a solution by making small improvements.
Positioning is used to find the next location following the simple steps of a rule.

Section 1. Number Patterns and Positions.

Example 1. Counting numbers and times tables.

Here are some number patterns that you need to know.
They are based on the times tables. Can you complete the table?

Sequence	What are the next 2?	What do we call these numbers?
1,2,3,4,5....	.... and	Counting numbers
2,4,6,8,10....	.... and	2 times table
3,6,9,12,15....	.... and	3 times table
4,8,12,16,20.....	.... and	
5,10,15,20,25.....	.... and	

Fill in these.

Example 2. Odd and even number.

ODD 1 3 5 7 9

EVEN 2 4 6 8 10

Example 3. Special sequences .

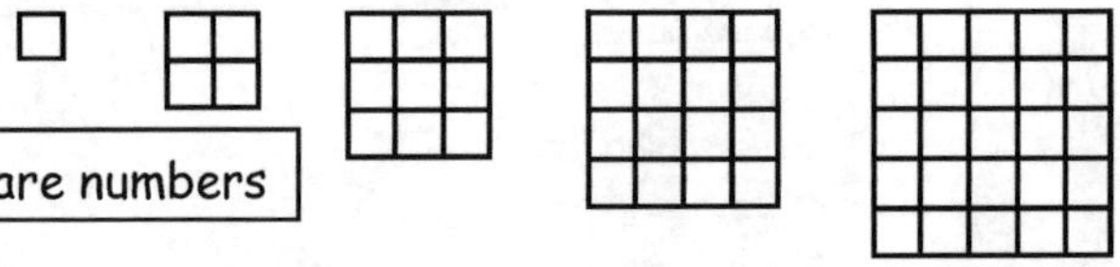

Square numbers

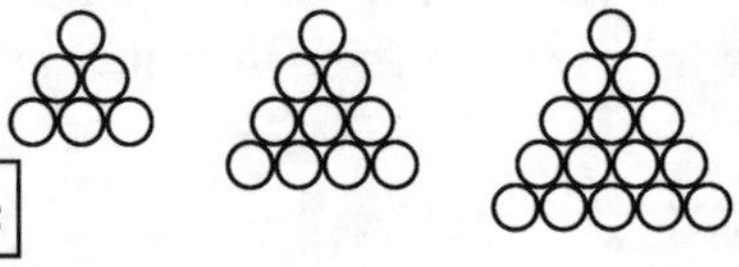

Triangular numbers

Both odd and even numbers go up in 2's. Odd numbers start at 1 and even numbers start at 2.
What do you notice about the number of small squares in each diagram?
The difference between each of the squares go up in odd numbers. The differences for the triangles go up in
then 3, then 4 etc...

Example 4. Sequences from patterns.

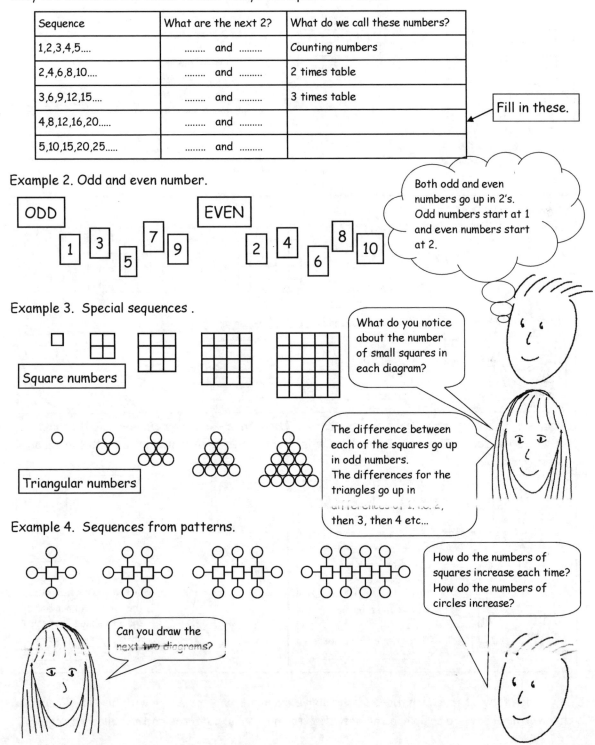

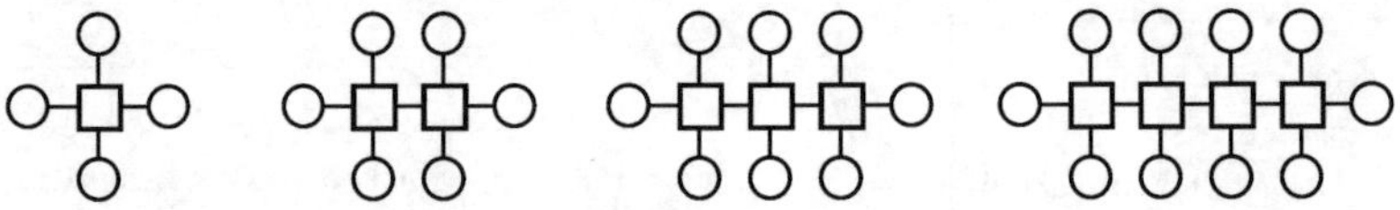

Can you draw the next two diagrams?

How do the numbers of squares increase each time? How do the numbers of circles increase?

How does this help us to improve non verbal reasoning?

Here are the Straight A Students who are going to show you what to look for when solving non verbal reasoning problems.

Sequence 1. Here is a sequence of grid. How would you describe the pattern formed?

They are also in a certain position so that the shaded squares increase along the top row and down the left column.
The first grid has 1 square shaded. The next has 3 shaded and then 5 and finally 7. this is the sequence of odd numbers.

Sequence 2. Look at the grids to the right here.
Can you explain how the pattern grows?

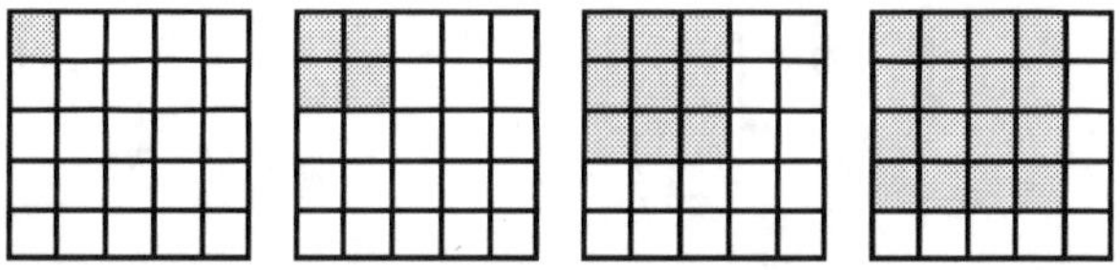

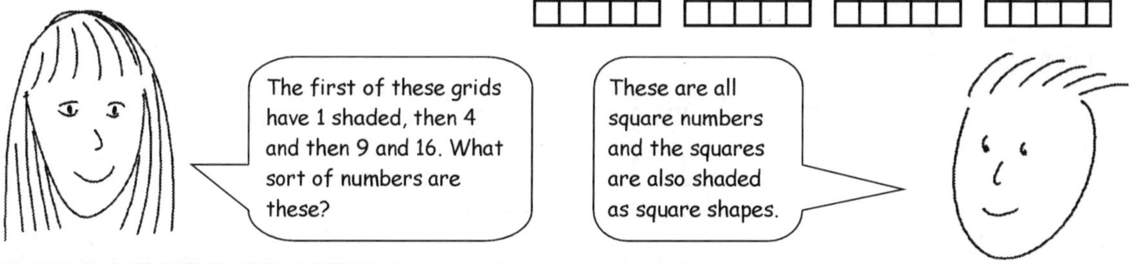

The first of these grids have 1 shaded, then 4 and then 9 and 16. What sort of numbers are these?
These are all square numbers and the squares are also shaded as square shapes.

Sequence 3. How does the first shape transform into the second? Then the third. Here is how it was explained.

Opposite corners of these grids are shaded. Two in the first, 4 in the second and 6 in the third.
These are even numbers.

Sequence 4. can you explain how the first shape has been transformed into the second?

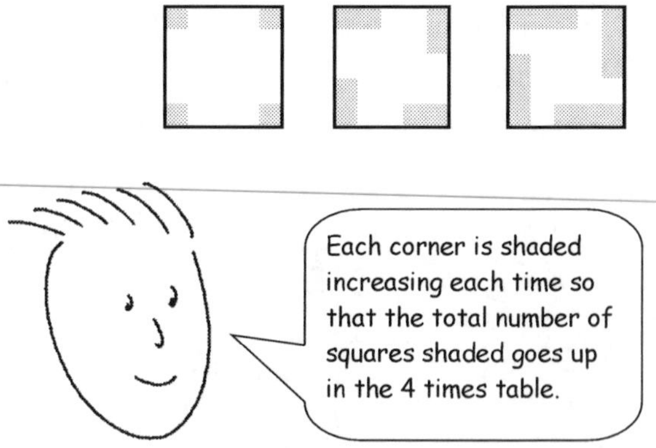

Each corner is shaded increasing each time so that the total number of squares shaded goes up in the 4 times table.

If you can follow the explanations of the above examples and you agree with how they continue then you are ready to complete the activities for numerical patterns on the following page.

Section 1. Number Patterns and Positions.

Numerical/Position Practice 1. Continue the following patterns by drawing the next diagram.

Pattern 1	Pattern 2	Pattern 3	Pattern 4	Pattern 5

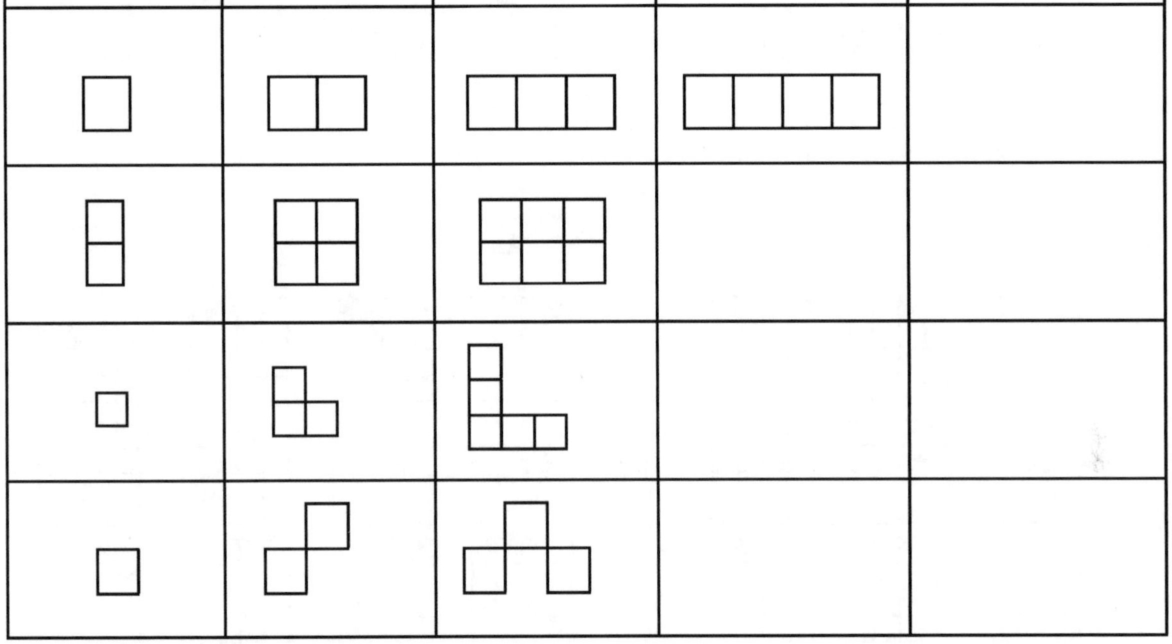

Numerical/Position Practice 2. There is a pattern in each of the rows of grids. Shade in the forth to complete the pattern.

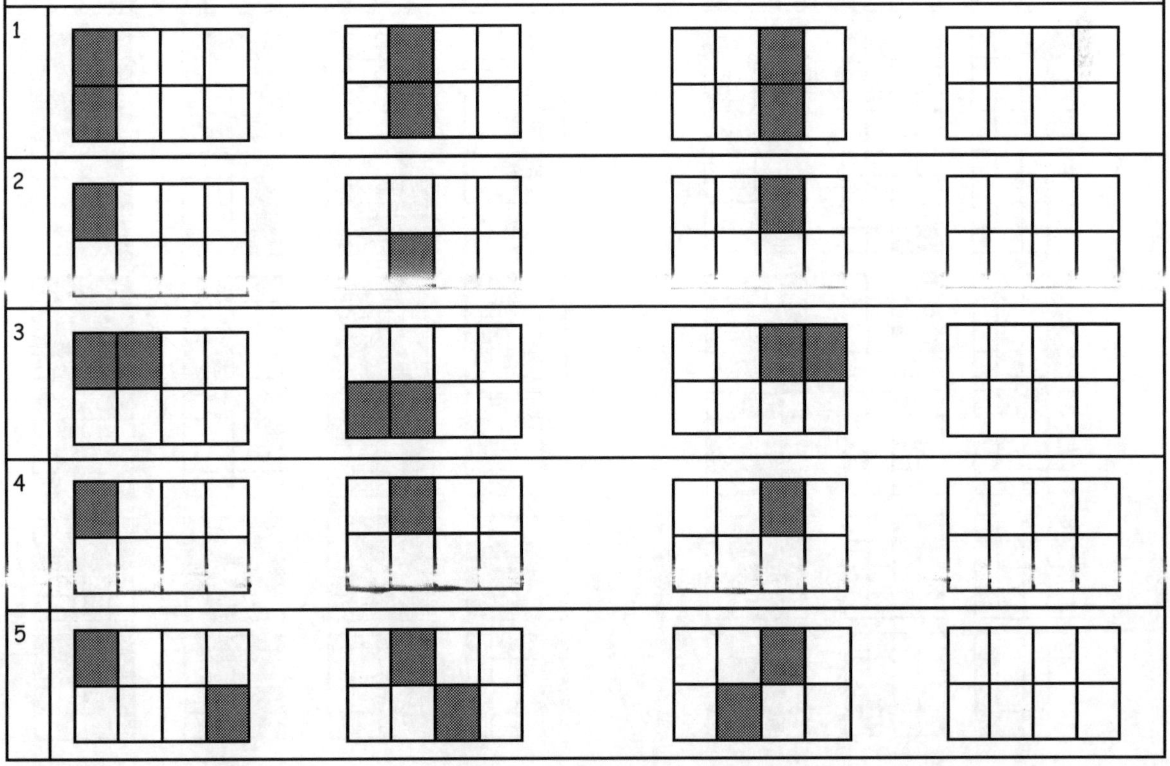

Numerical Worksheet 1. Growing patterns in a grid.

Look at the sequence of patterns formed in the first 3 grids and decide how they change from the left grid to the right grid. Then find which of A,B,C and D that best continues the pattern.

Example: Complete the sequence. A B C D

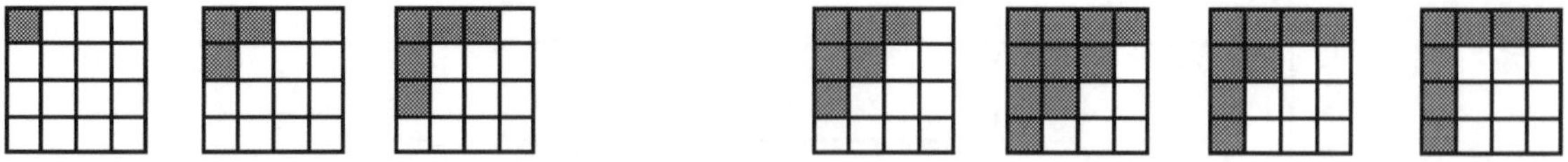

The numbers of shaded squares goes 1, then 3 and then 5. There location is top left corner increasing along and down by 1 square from one grid to the next. The answer you should get is D.

Complete the sequence. A B C D

1

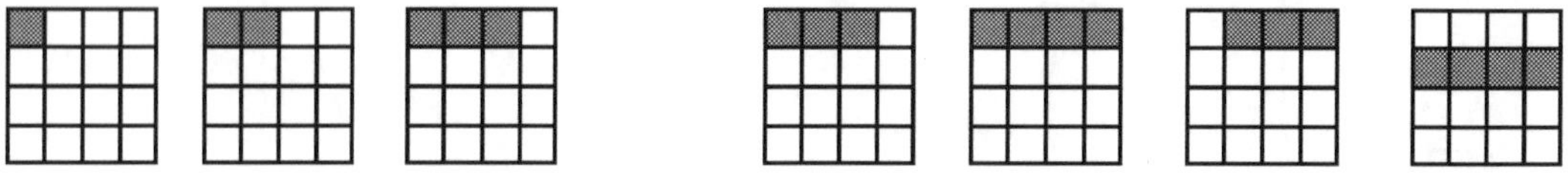

2

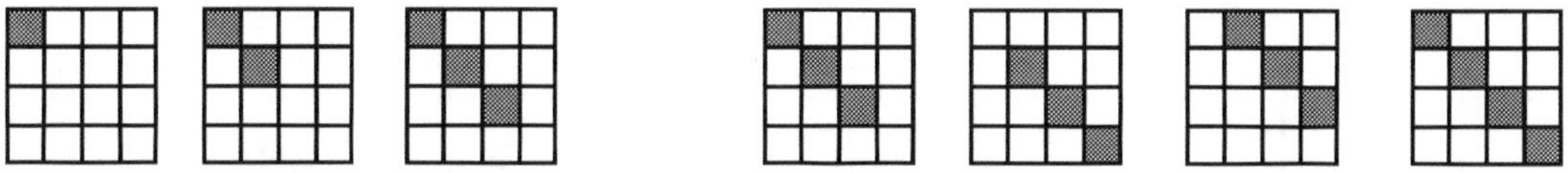

3

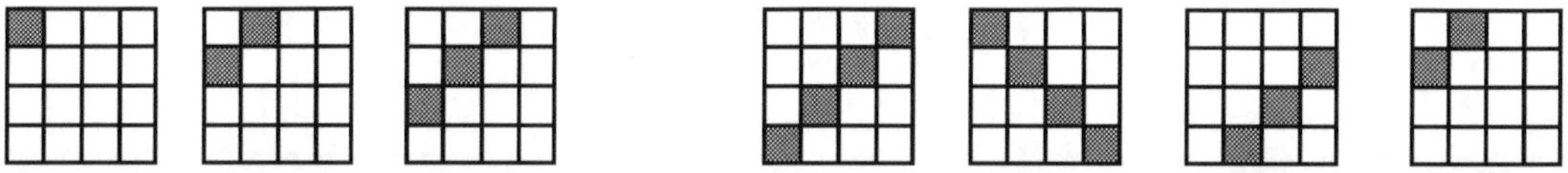

4

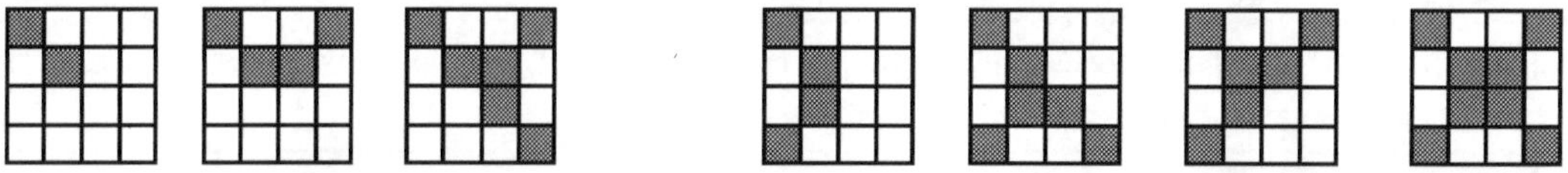

5

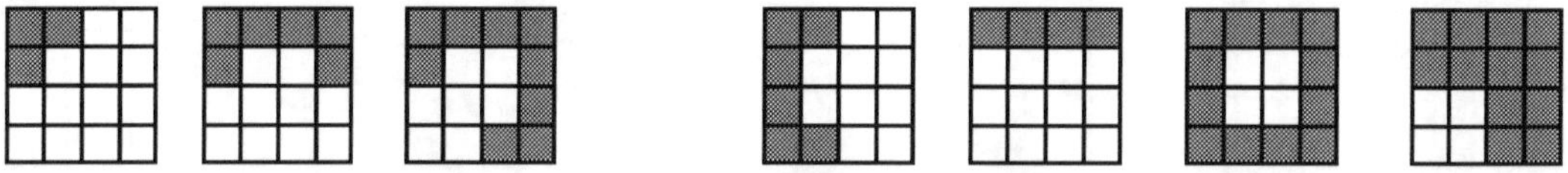

6

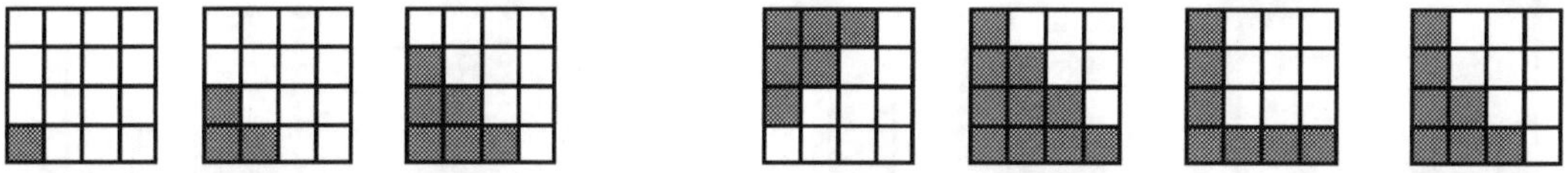

7

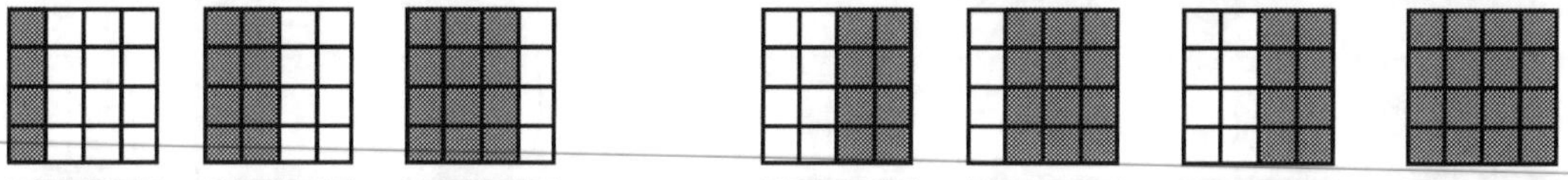

8

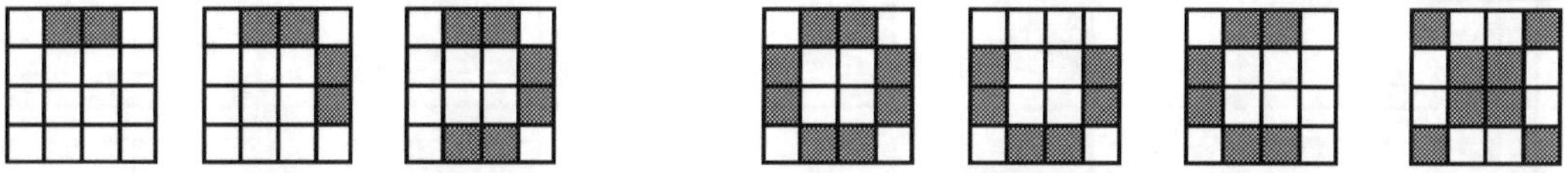

9

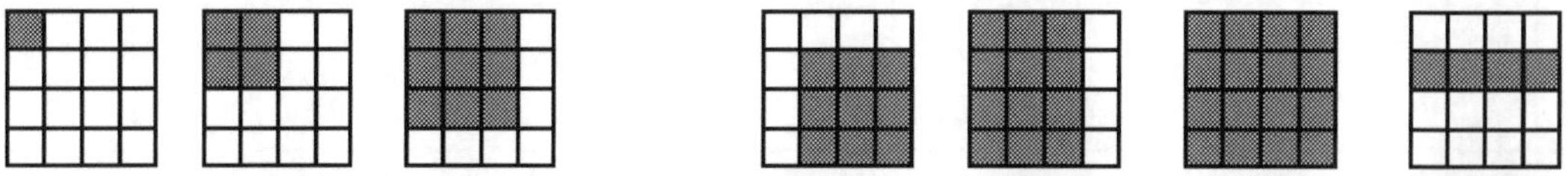

10

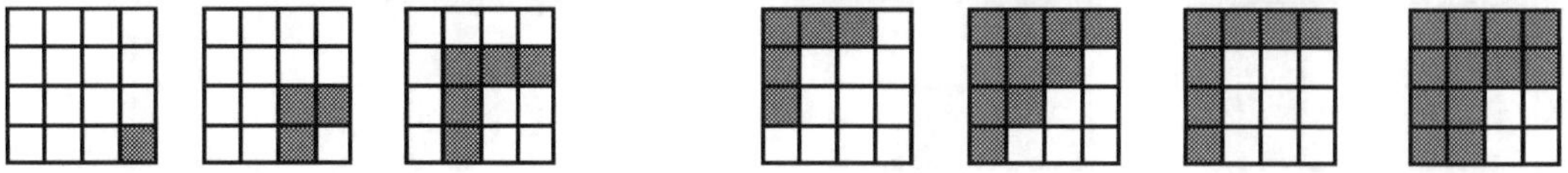

Numerical Worksheet 2. Growing patterns in a grid.

Look at the sequence of patterns and decide how they change from the left grid to the right grid A,B,C,D and E that completes the square.

Example. Complete the sequence.

A B C D

This pattern starts in grid 1 with 2 squares in opposite corners shaded. The pattern changes from grid to grid by increasing outwards so that the total increases by 4 each time. The answer you should get is B.

Complete the sequence.

A B C D

1 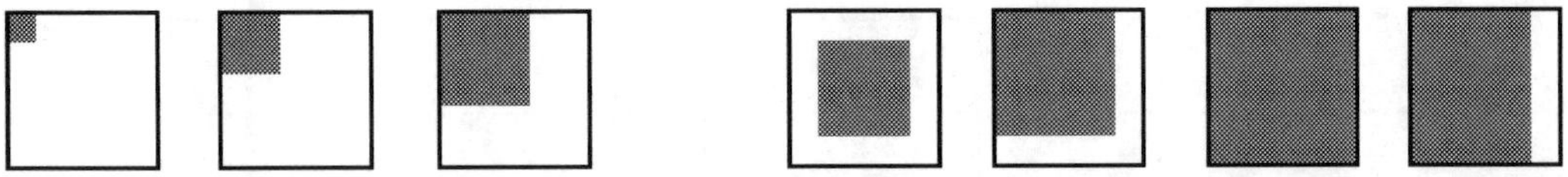

2 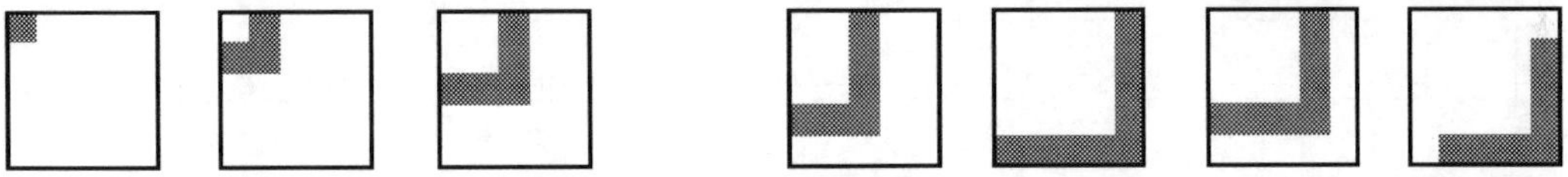

3 

4 

5 

6 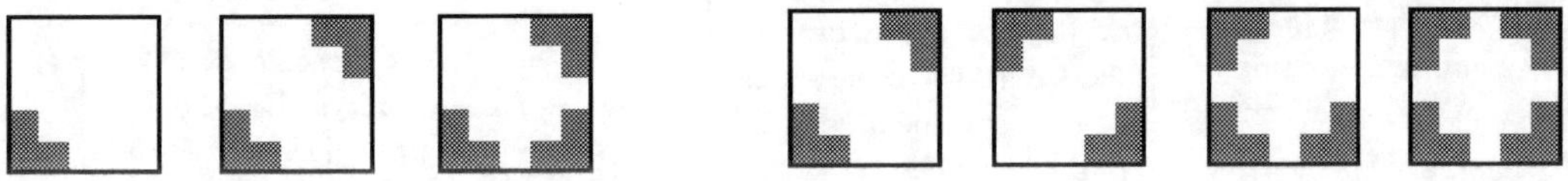

7 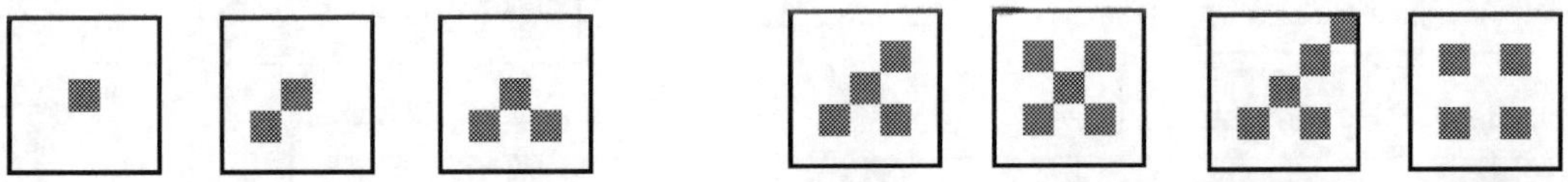

8 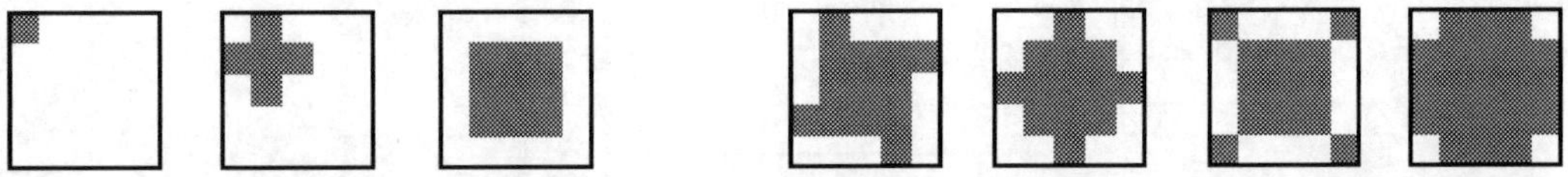

9 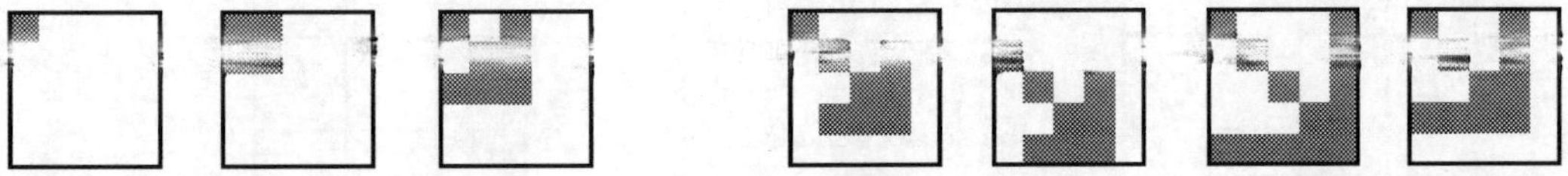

10 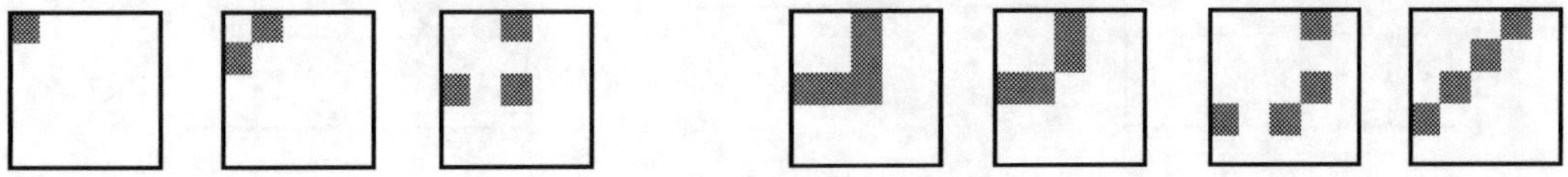

Section 2. Reflective Symmetry.

Example 1. Mirror lines or lines of symmetry.

Complete each of the drawings so that the dashed line is a line of symmetry.

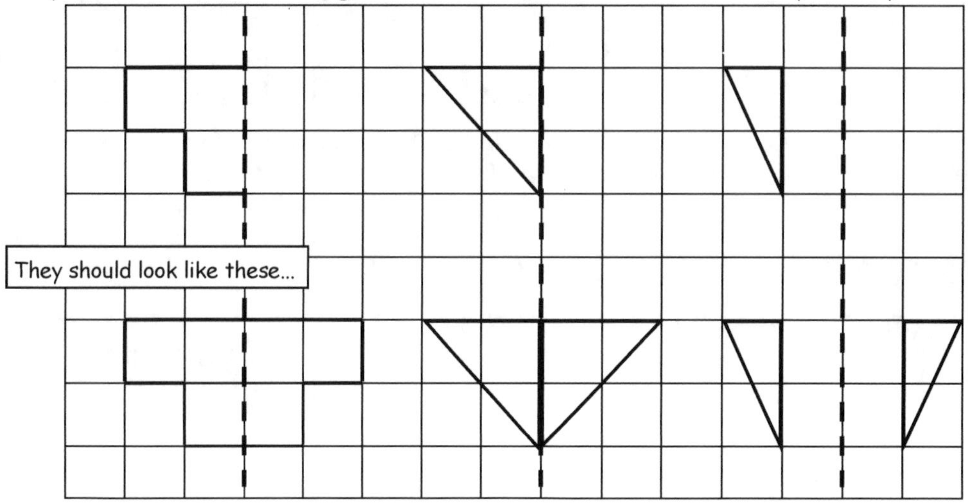
They should look like these...

Example 2. Symmetry on a square grid.

Shade in the squares in the grid so the dashed line is a line of symmetry.

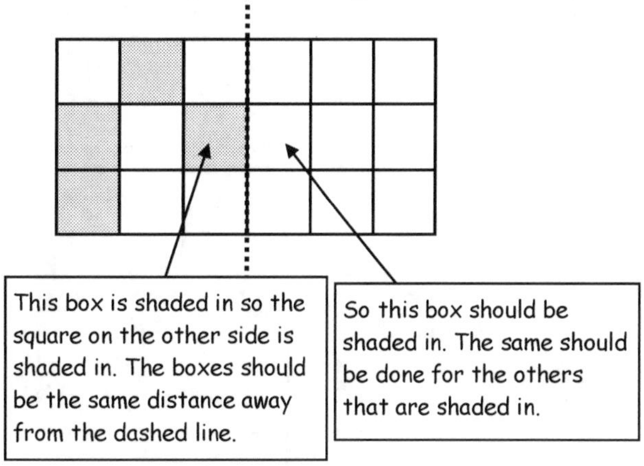
This box is shaded in so the square on the other side is shaded in. The boxes should be the same distance away from the dashed line.
So this box should be shaded in. The same should be done for the others that are shaded in.

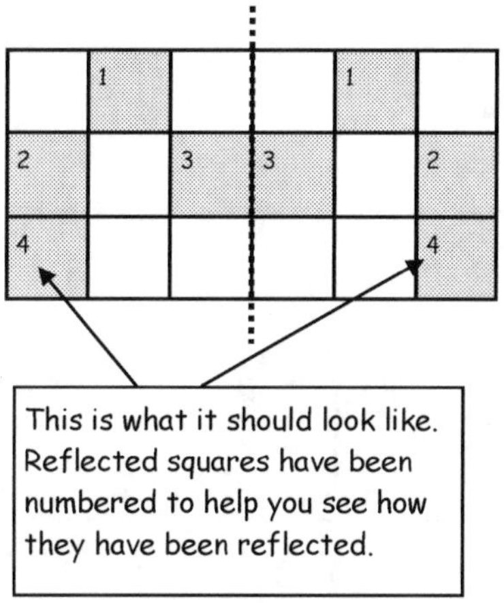
1
1
2
3
3
2
4
4
This is what it should look like. Reflected squares have been numbered to help you see how they have been reflected.

Example 3. Diagonal mirror lines.

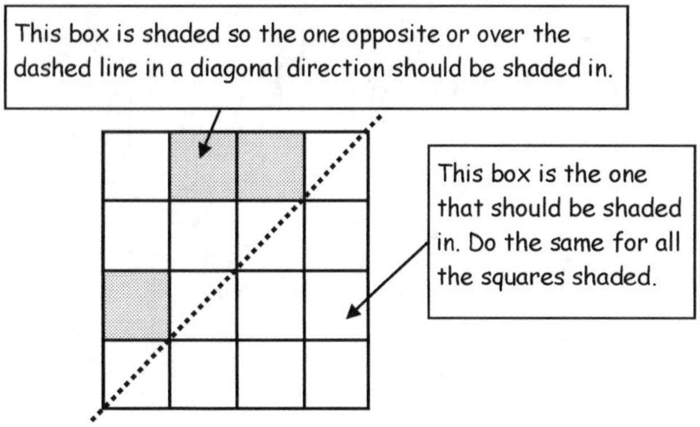
This box is shaded so the one opposite or over the dashed line in a diagonal direction should be shaded in.
This box is the one that should be shaded in. Do the same for all the squares shaded.

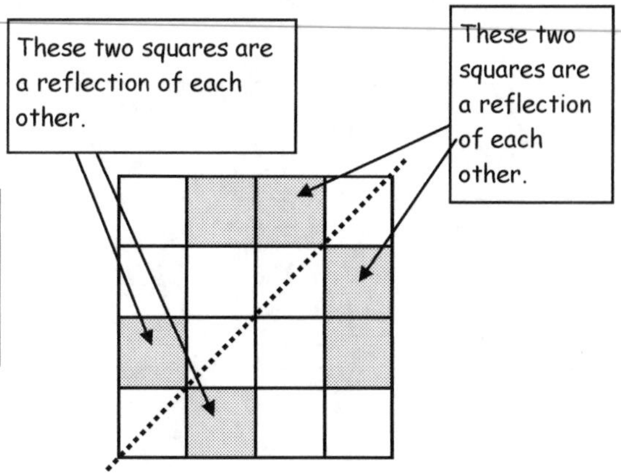
These two squares are a reflection of each other.
These two squares are a reflection of each other.

How does reflection symmetry help to improve non verbal reasoning?

Here are the Straight A Students who are going to show you what to look for when solving non verbal reasoning problems.

Example 1. Here is a grid with a dashed mirror line. Which one of the three shows the true reflection after the shaded squares are reflected in the dashed mirror line?

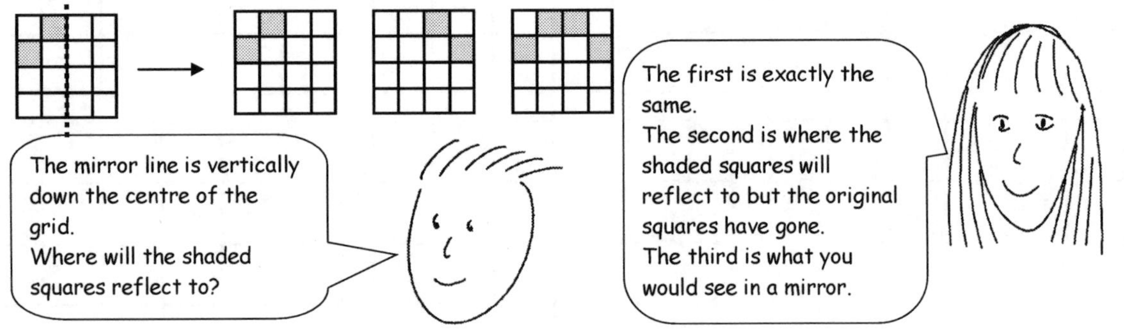
The mirror line is vertically down the centre of the grid.
Where will the shaded squares reflect to?
The first is exactly the same.
The second is where the shaded squares will reflect to but the original squares have gone.
The third is what you would see in a mirror.

Sequence 2. Look at the grids to the right here.
Can you explain how the pattern grows?

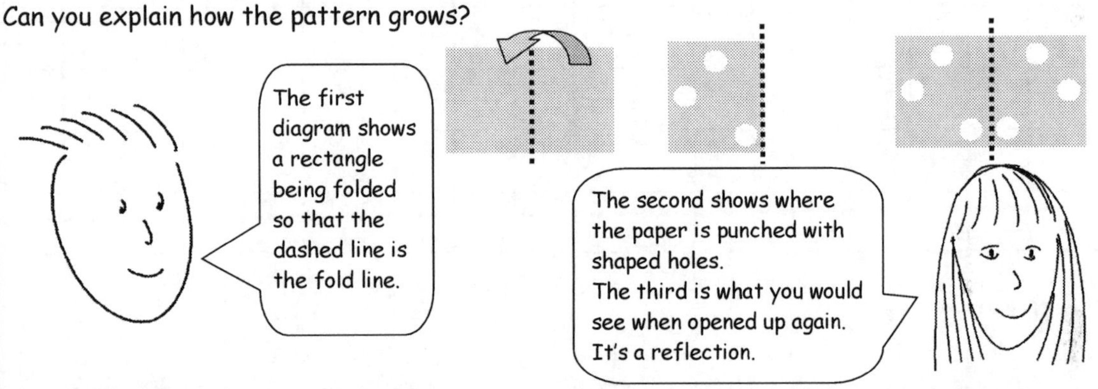
The first diagram shows a rectangle being folded so that the dashed line is the fold line.
The second shows where the paper is punched with shaped holes.
The third is what you would see when opened up again.
It's a reflection.

Example 3. How does the first shape transform into the second? Here is how it was explained.

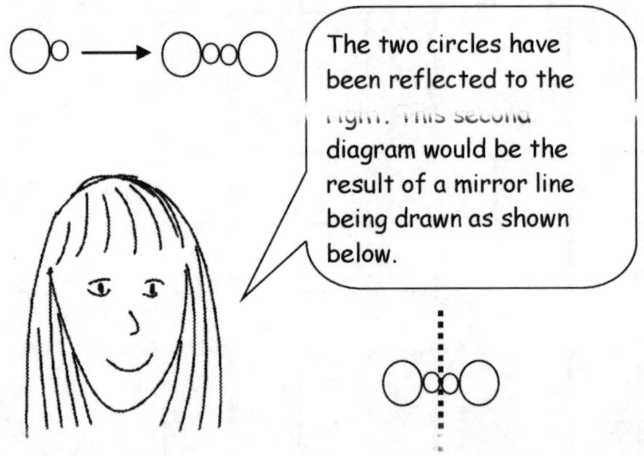
The two circles have been reflected to the
diagram would be the result of a mirror line being drawn as shown below.

Example 4. Can you explain how the first shape has been transformed into the second?

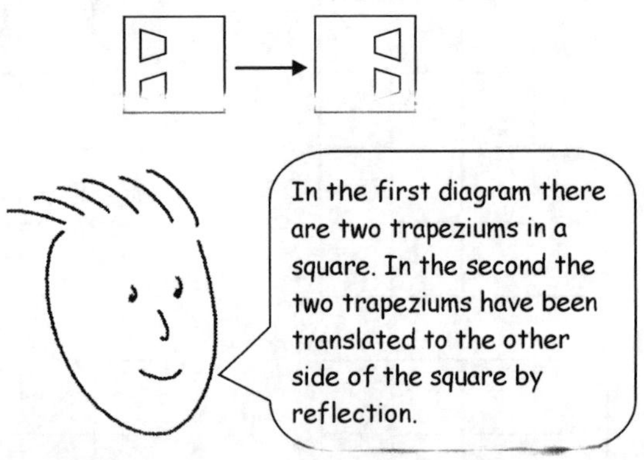
In the first diagram there are two trapeziums in a square. In the second the two trapeziums have been translated to the other side of the square by reflection.

If you can follow the explanations of the above examples and you agree with how they continue then you are ready to complete the activities for reflection symmetry on the following page.

Reflection Practice 1.

Shade in the squares on the right hand side only to make the dotted line a line of symmetry. The first one has been done for you.

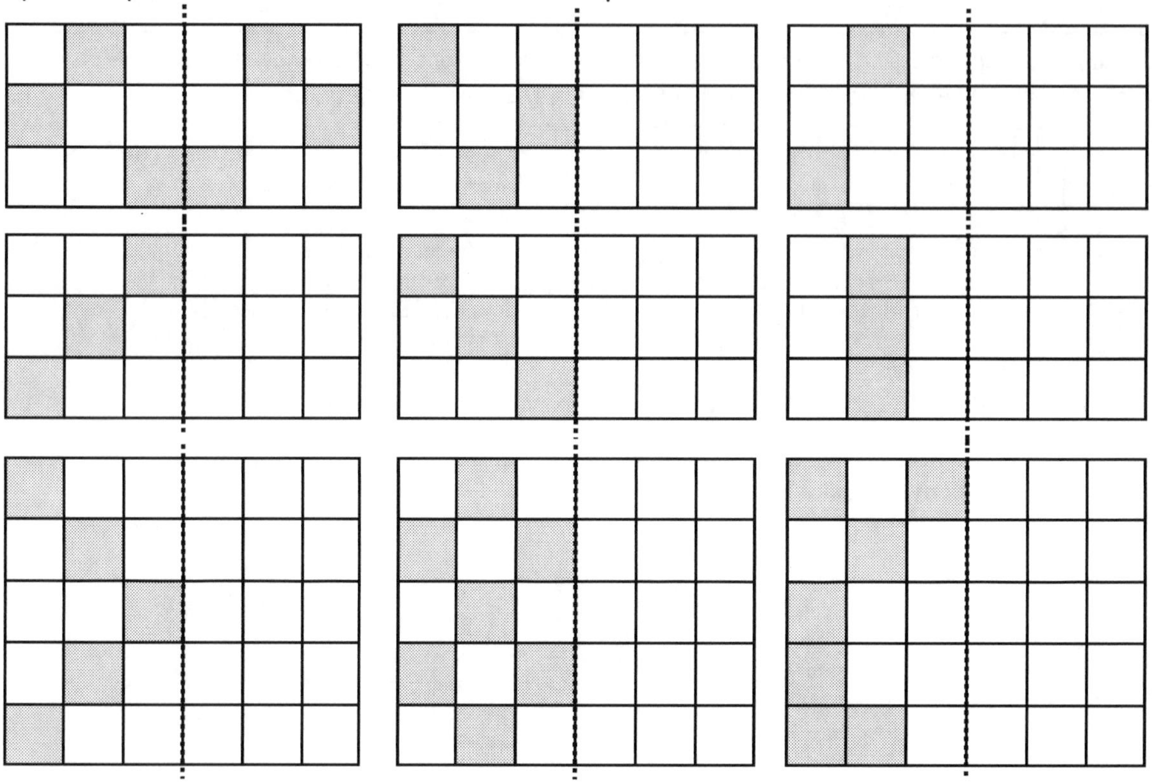

Reflection Practice 2.

Shade in the squares on the other side of the diagonal side to make the dotted line a line of symmetry.

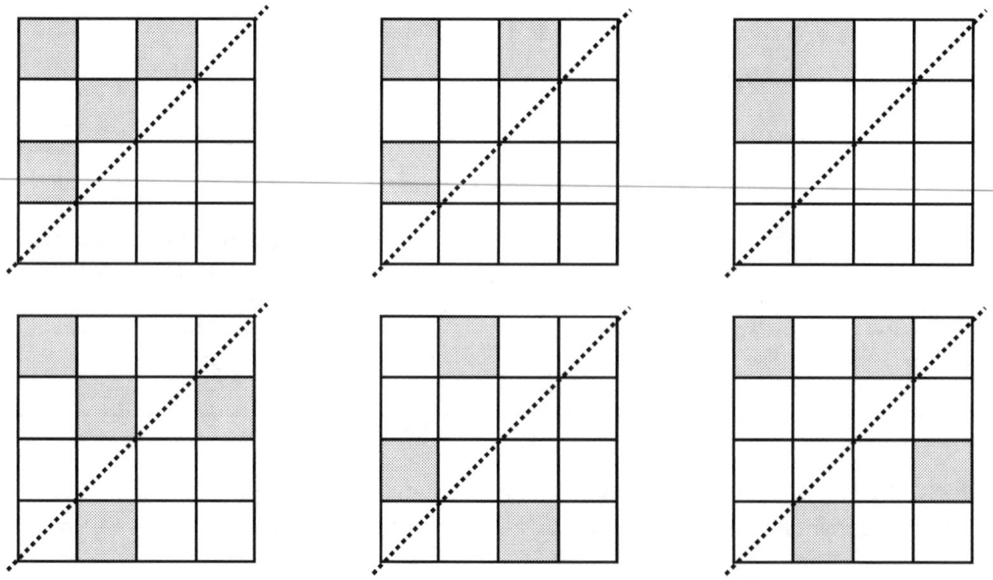

Reflection Practice 3.

Shade in the squares in the other three parts to complete the symmetry pattern.
The example has been done for you.

Example Step 1 Step 2 Step 3

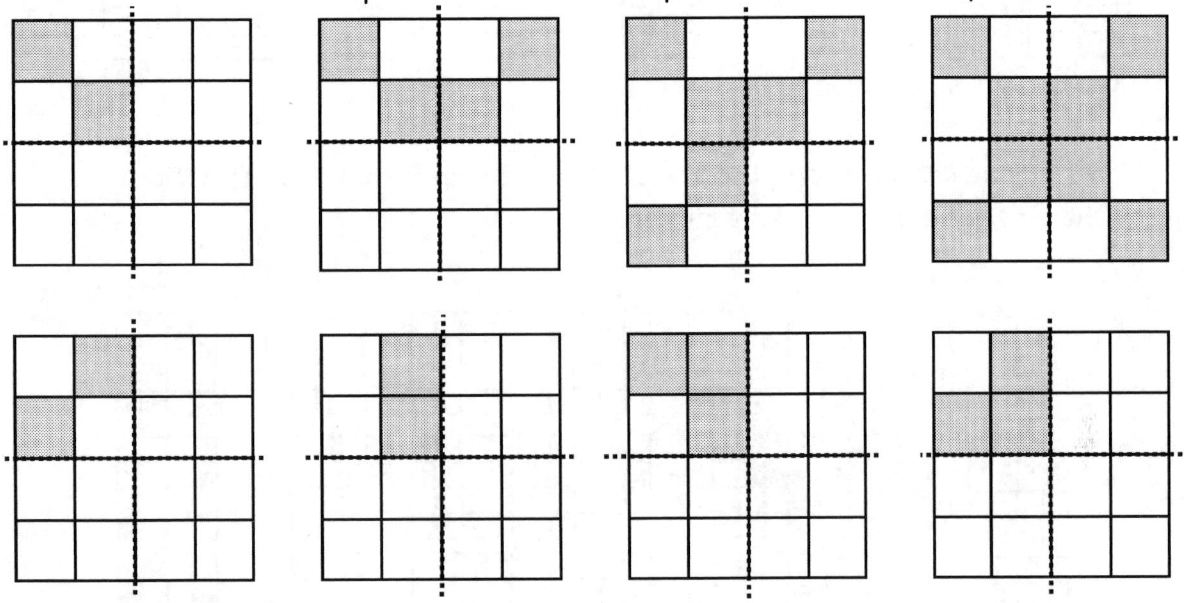

Reflection Practice 4.

Shade in the squares in the other three parts to complete the symmetry pattern.

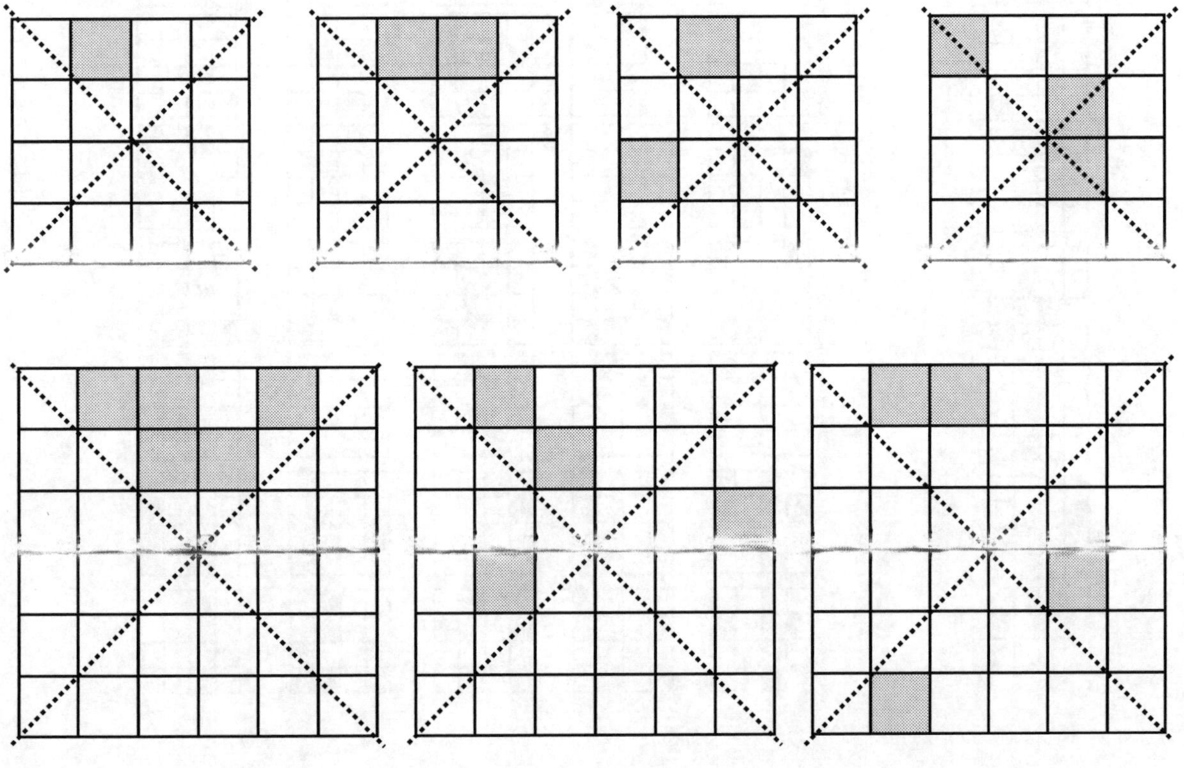

Reflection Worksheet 1. Reflecting patterns in a mirror line.

Find the square grid from the choices A,B,C,D and E that is a reflection of the first grid in the dashed mirror line.

Example: Find the reflection. A B C D E

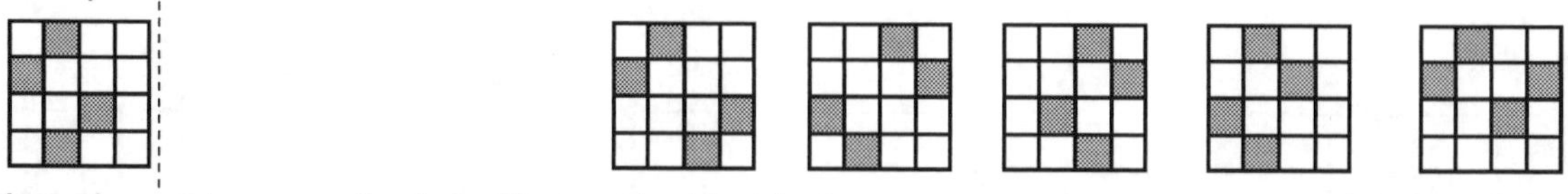

If you place a mirror over the dashed line you should get is C.

Now find the square grid in the choices A,B,C,D and E that is a reflection of the first grid in the dashed mirror line.

Find the reflection. A B C D E

1
2
3
4
5
6
7
8
9
10

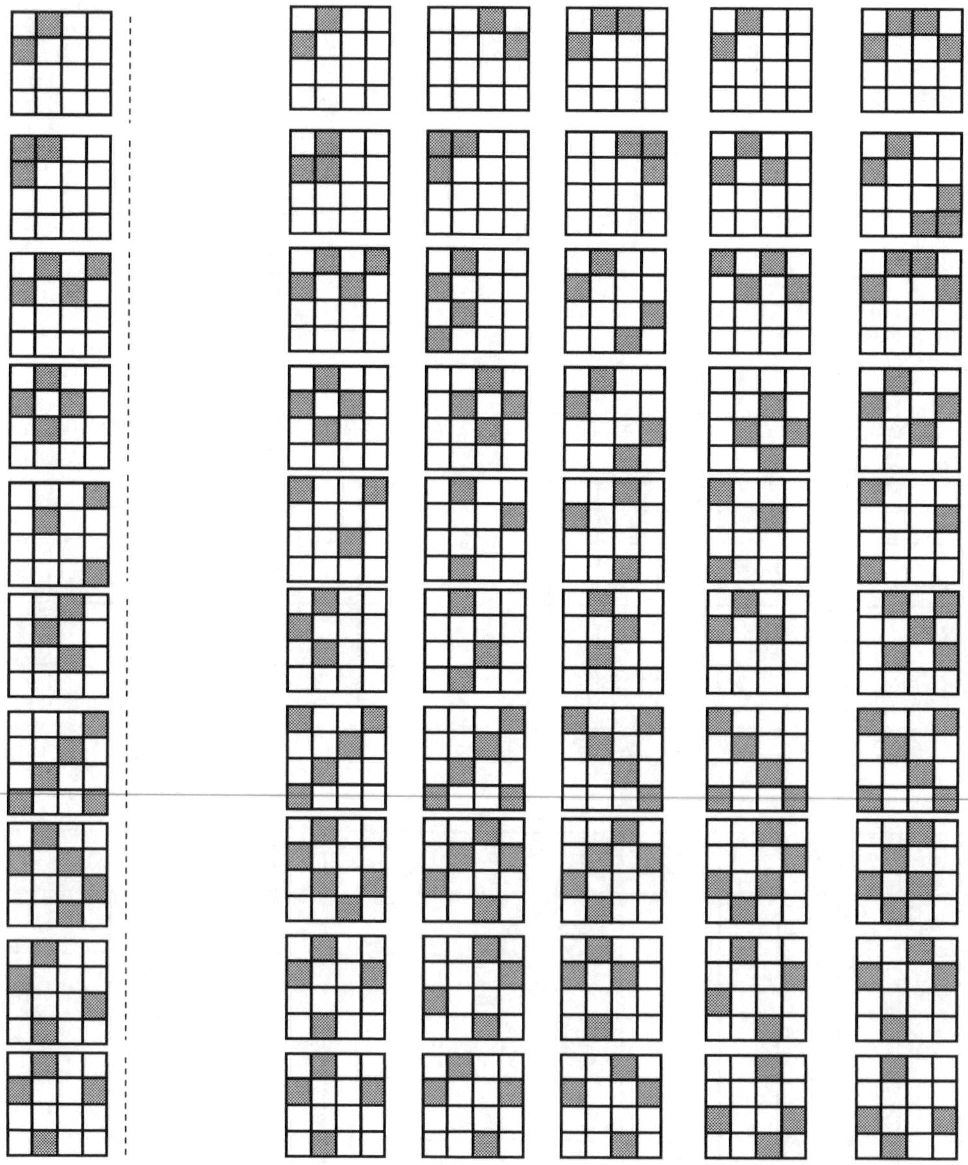

Reflection Worksheet 2. Reflecting patterns in a mirror line.

Find the grid from the choices A,B,C,D and E that completes the square grid if the dashed line is a line of reflection or mirror line.

Example: Find the reflection.

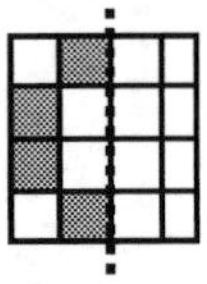

A B C D E

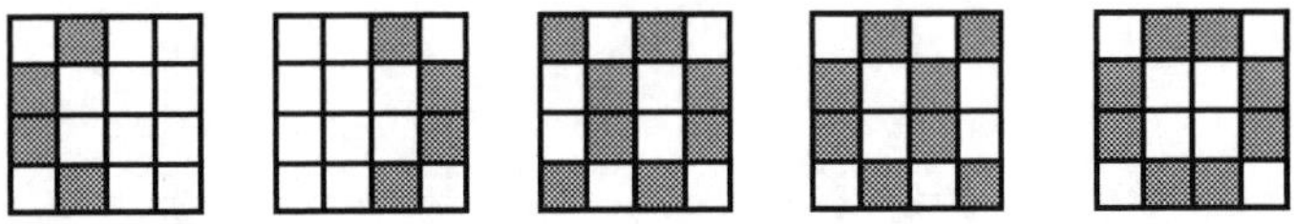

Using laws of reflection or a mirror placed over the dashed line you should get E.

Find the reflection.

A B C D E

1

2

3

4

5

6

7

8

9

10

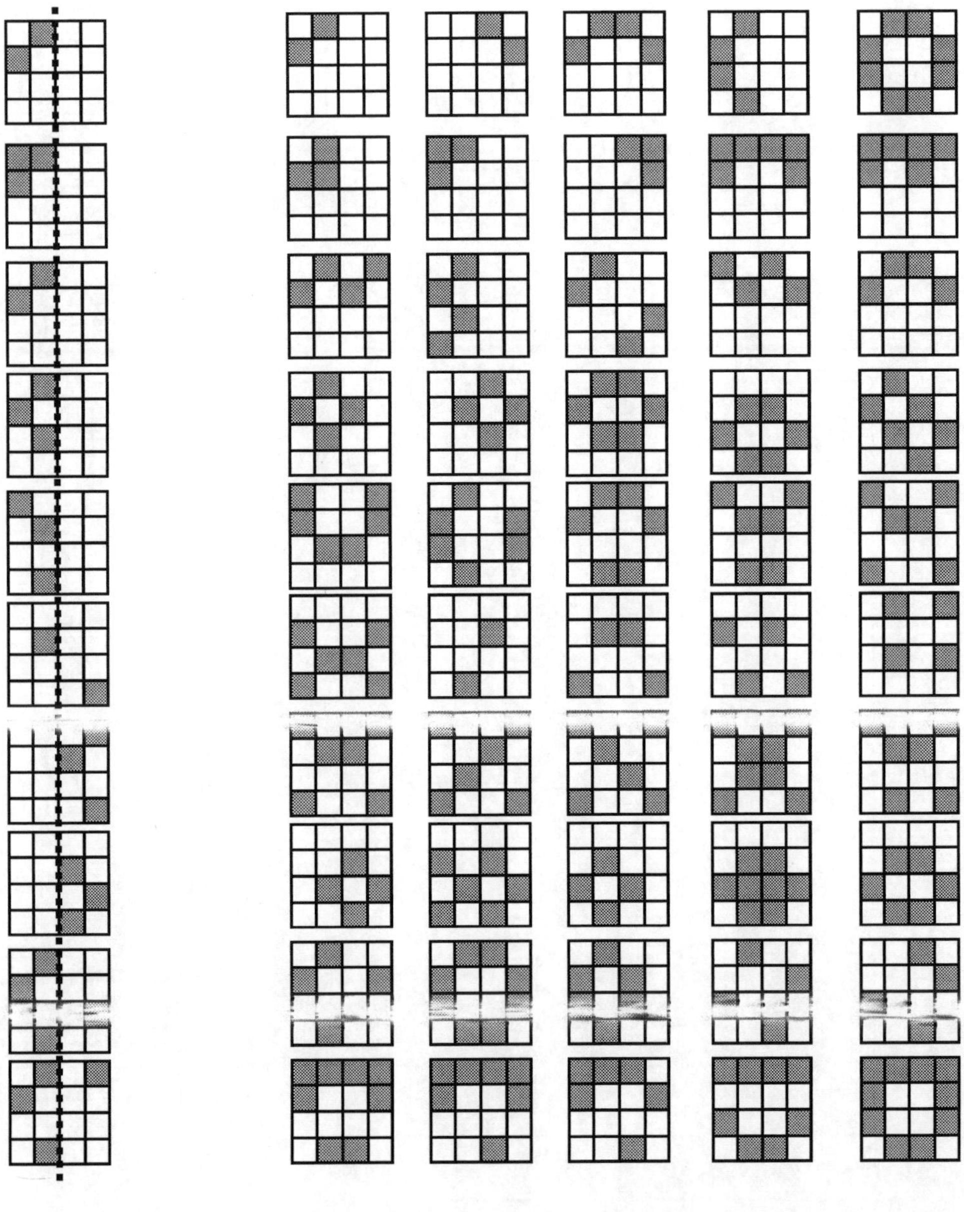

Reflection Worksheet 3. Reflecting patterns in a mirror line.

Here the shape next to the dashed mirror line is reflected. You need to find one of the shapes labelled A,B, C, D or E which is its reflection.

Example: Find the reflection. A B C D E

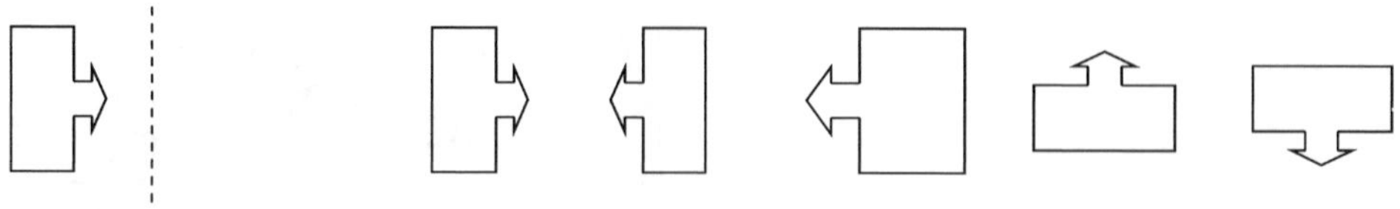

This is a simple reflection in the mirror line shown. You would find a mirror useful and you should get shape B.

Find the reflection. A B C D E

Reflection Worksheet 4. Folding paper patterns (1 fold).

Find the grid from the choices A,B,C and D that will be the result of folding the paper once as shown and then punching a shaped hole through both sides of the fold.

Example. Folding and hole punching.

A B C D

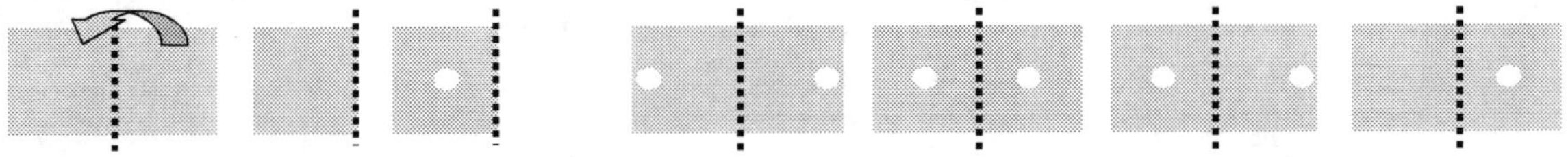

The hole is punched in the centre of the folded paper so that when the paper is unfolded it appears as if it has been reflected over the fold line. The answer you should get is B.

Now try these

A B C D

1
2
3
4
5
6
7
8
9
10

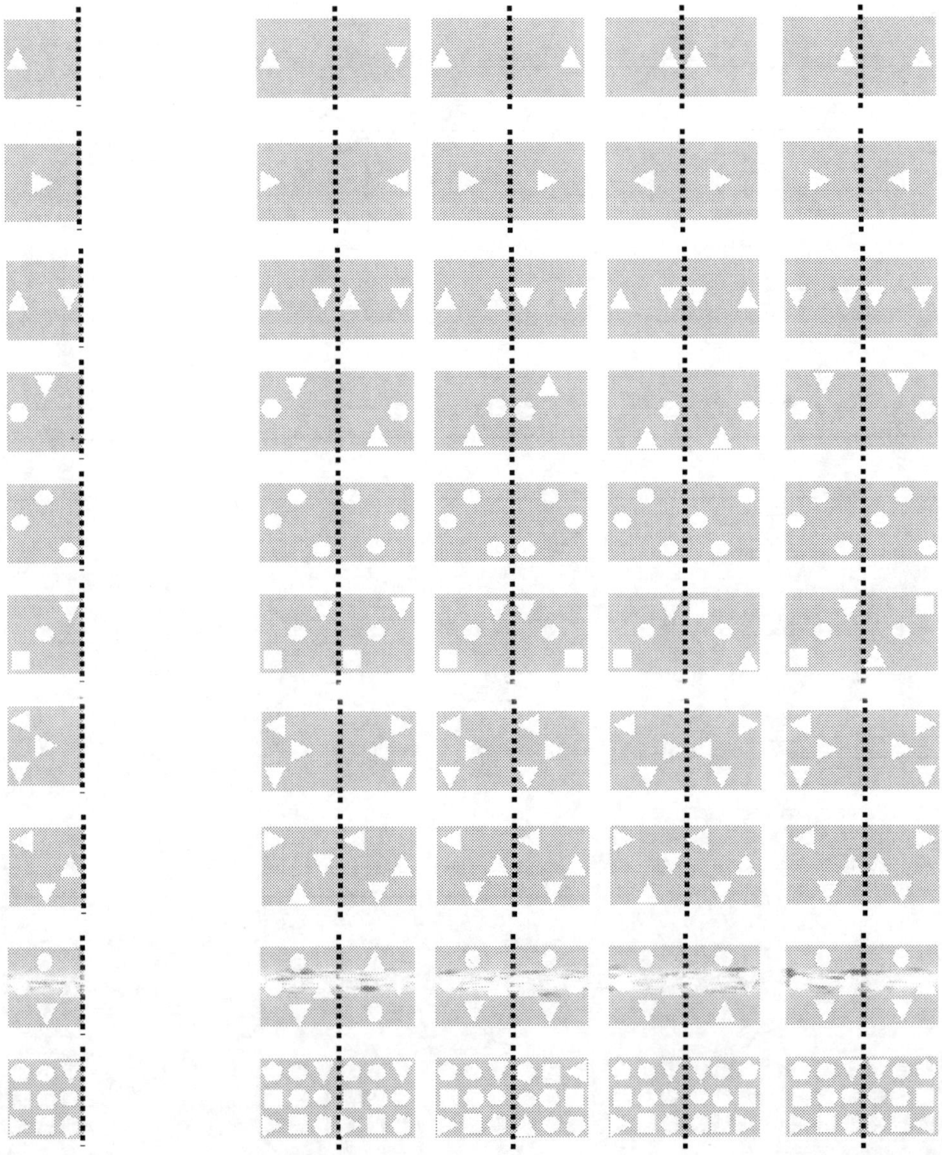

Reflection Worksheet 5. Reflecting patterns in a mirror line.

Here the first shape changes into the second shape using reflection in a mirror line. You need to find one of the shapes A,B, C or D which follows the same change.

Example. Find the reflection. A B C D

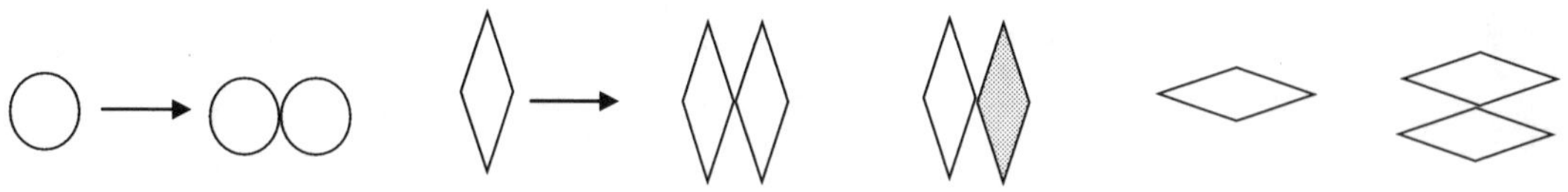

The circle has changed into 2 circles as if there is a mirror line next to the first so that it is reflected. You should get shape A.

Find the reflection. A B C D

1
2
3
4
5
6
7
8
9
10

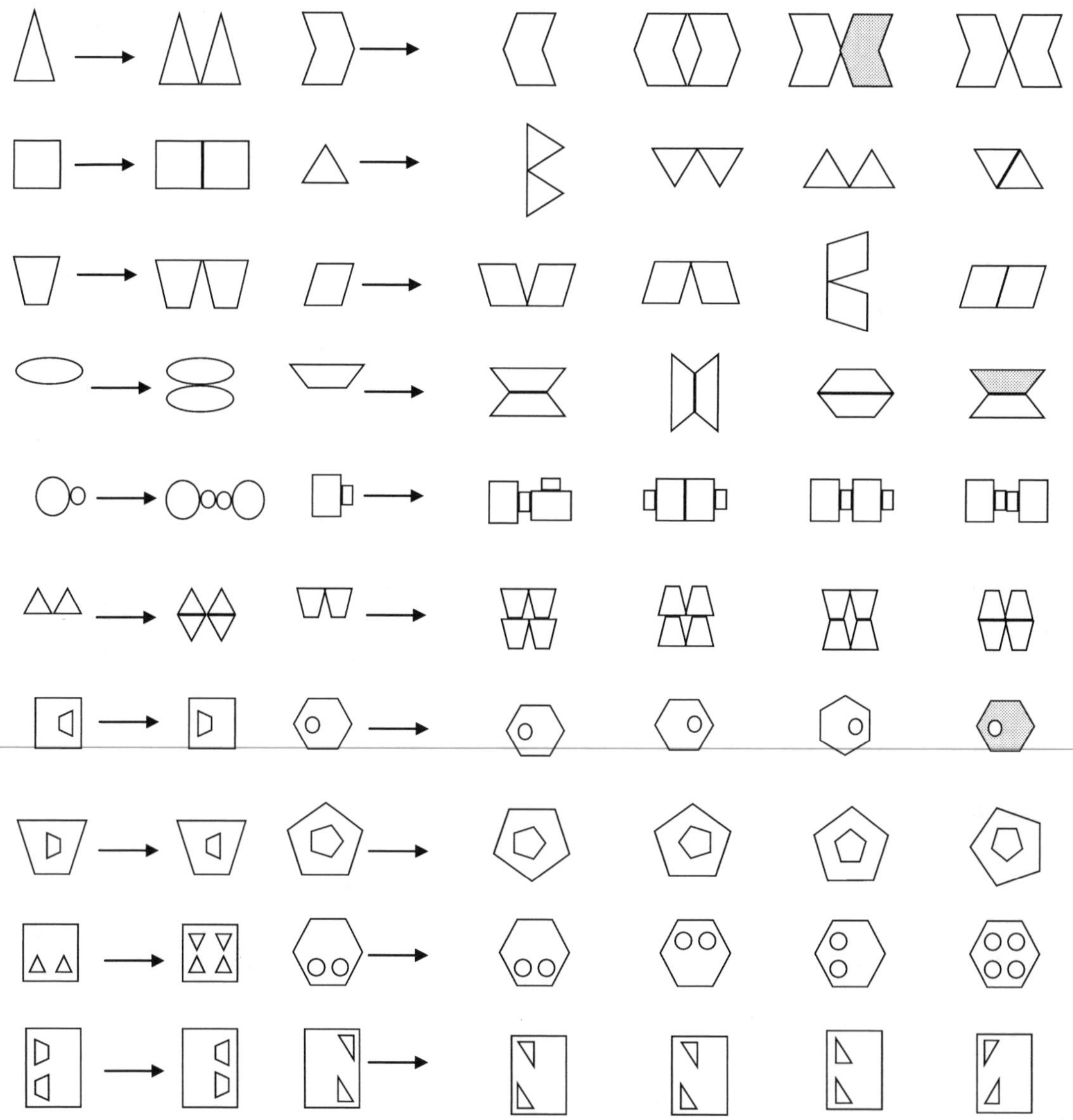

Section 3. Rotational Symmetry.

Example 1. Finding shapes with rotational symmetry.

All these are Greek letter and they have also something else in common. Can you see what it is? If you are not sure try turning them upside down. What do you notice?

Θ Η Φ Ξ Χ Ζ

Trace the shape with tracing paper. You can also use baking sheets or grease proof paper!! It's just as good... Then turn the sheet upside down. Does it fit the over the shape exactly?

Each one is the same upside down. This means it looks the same in 2 positions.

We say all these shapes have a rotational symmetry.

Looks the same in 2 positions means rotational symmetry **order 2.**

Example 2. Completing a grid with rotational symmetry of order 2.

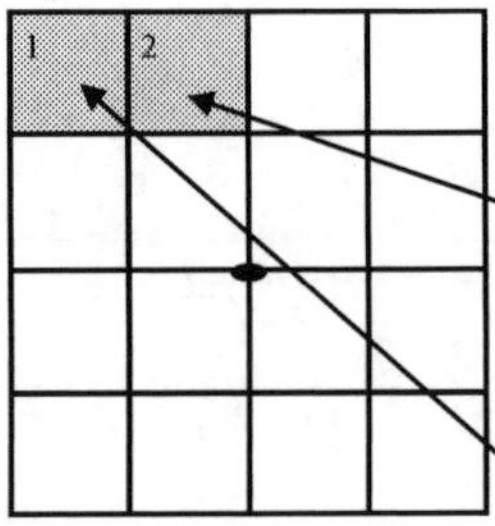

Trace and shade the squares.
Turn the tracing paper upside down to find which of squares must be shaded.

If this square(2) is shaded then this square should be shade as well. Labelled (2).

If this is shaded (1) then this square should be shaded too. Labelled (1).

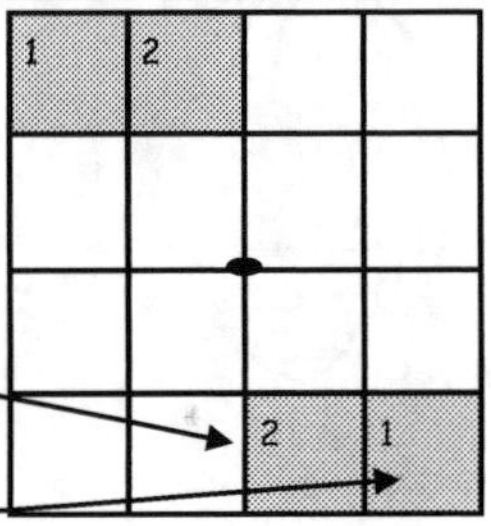

Example 3. Shapes with rotational symmetry order 3.

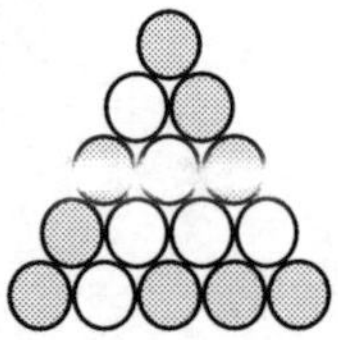

Trace the triangular shape and rotate to find how many positions it will look exactly the same with one full turn of the tracing paper in the positions shaded.

3 positions means rotational symmetry of order 3.

Example 4. Square grid with rotational symmetry order 4.

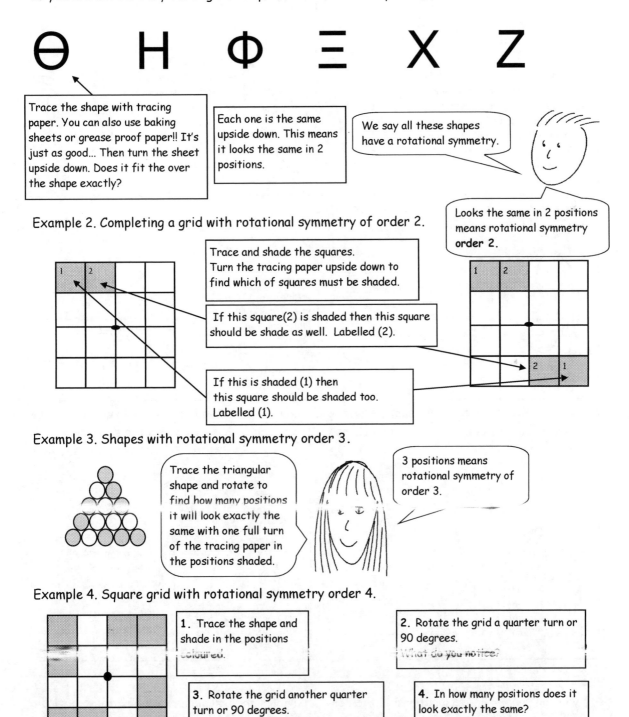

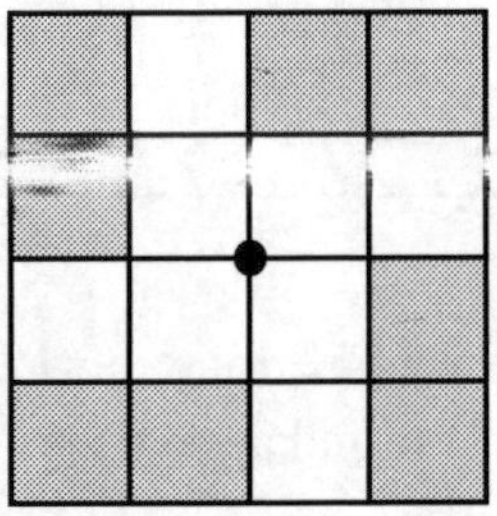

1. Trace the shape and shade in the positions coloured.

2. Rotate the grid a quarter turn or 90 degrees.
What do you notice?

3. Rotate the grid another quarter turn or 90 degrees.
What do you notice? Repeat until the grid is back to the start.

4. In how many positions does it look exactly the same?

How does rotational symmetry help us to improve non verbal reasoning?

Here are the Straight A Students who are going to explain what to look for when solving non verbal reasoning problems.

Sequence 1. Here is a sequence of grid. How would you describe the pattern formed?

Each shape consists of a grid of 16 squares. In which 2 are squares shaded. The grid is rotated through ¼ turn clockwise about the point in the centre of the grid. This is how you get from one grid to the next.
Using tracing paper and the point of your pencil will help you find where the shaded squares move to. You can even use baking sheets!

Sequence 2. Look at the grids to the right here.
Can you explain how the pattern grows?

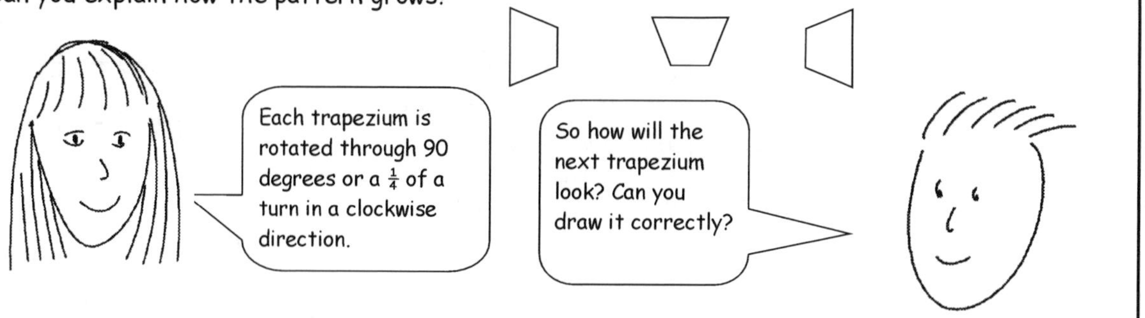
Each trapezium is rotated through 90 degrees or a ¼ of a turn in a clockwise direction.
So how will the next trapezium look? Can you draw it correctly?

Diagram 3. How does the first shape transform into the second? Here is how it was explained.

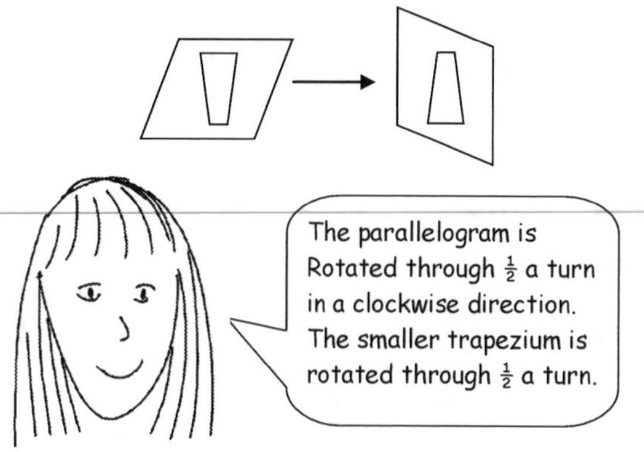
The parallelogram is Rotated through ½ a turn in a clockwise direction. The smaller trapezium is rotated through ½ a turn.

Diagram 4. can you explain how the first shape has been transformed into the second?

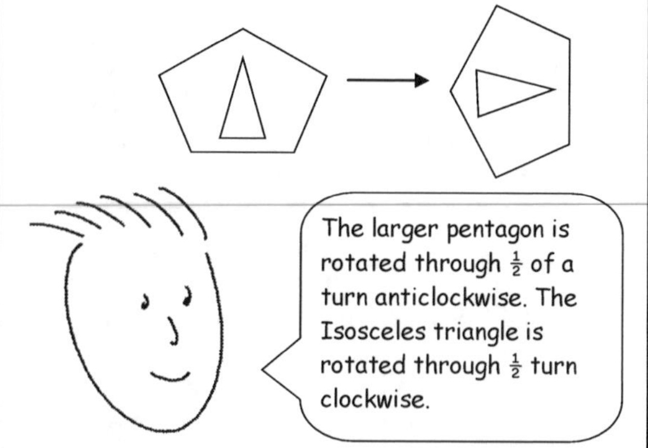
The larger pentagon is rotated through ½ of a turn anticlockwise. The Isosceles triangle is rotated through ½ turn clockwise.

If you can follow the explanations of the above examples and you agree with how they continue then you are ready to complete the activities for rotational symmetry on the following page.

Rotation Practice 1.

Here is the English alphabet. Put a circle around all the letters that look the same when turned upside down.

A B C D E F G H I

J K L M N O P Q R

S T U V W X Y Z

Rotation Practice 2.

Here is the Greek alphabet. Put a circle around all the letters that look the same when turned upside down.

A B Γ Δ E Z H Θ I

K Λ M N Ξ O Π P

Σ I Υ Φ X Ψ Ω

Rotation Practice 3.

Shade in the squares in the other three parts to complete the rotational symmetry pattern. The example has been done for you.

Example ⟶ Step 1 ⟶ Step 2 ⟶ Step 3

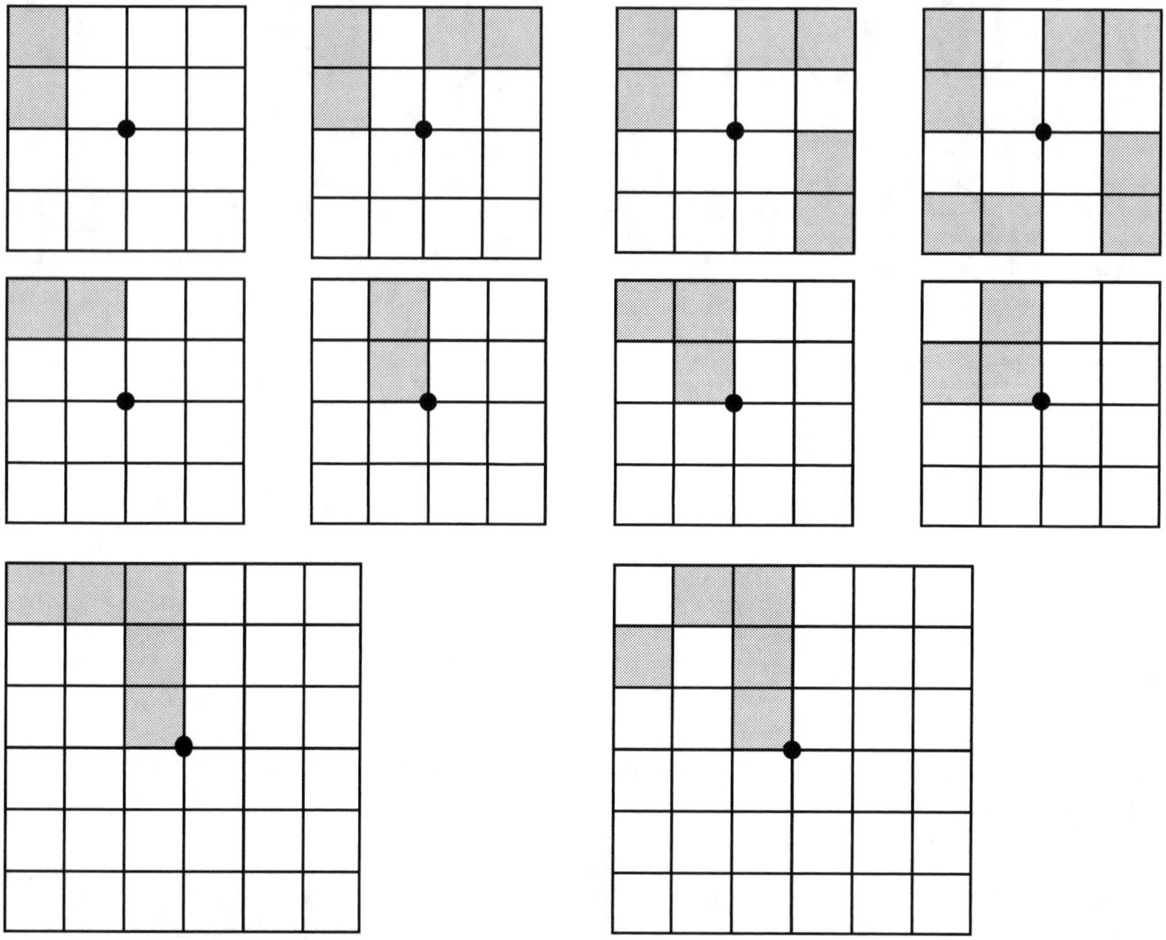

Rotation Practice 4.

In each of the triangles below shade in the least amount of circles to complete the rotational symmetry pattern of order 3.

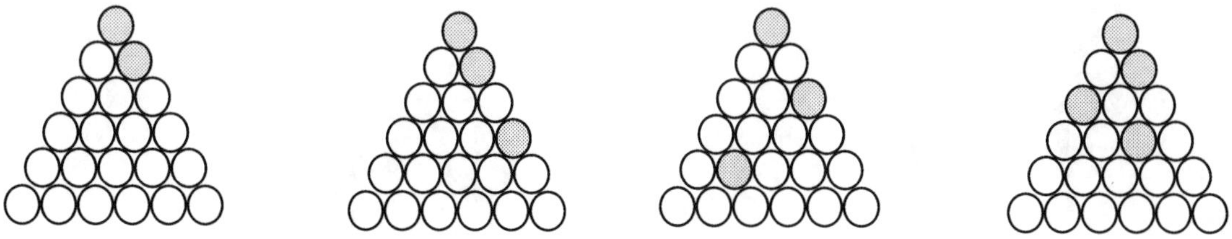

Rotation Worksheet 1. Shapes rotating in a sequence.

The first three shapes are all in a sequence of rotation. Find the fourth A, B, C or D that completes the sequence of rotational symmetry.

Example; First three shapes. A B C D

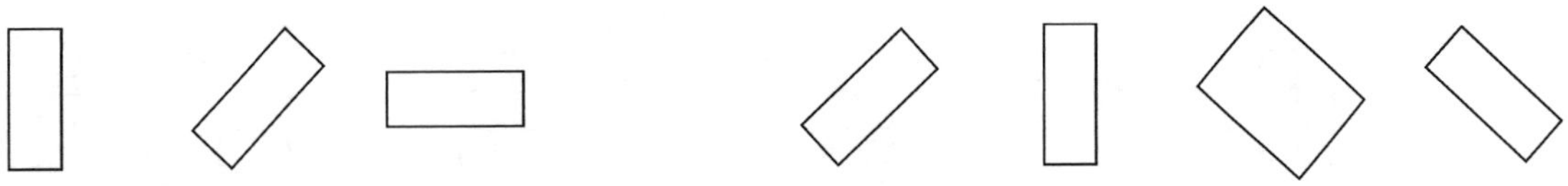

The sequence simply rotates in a clockwise direction. The answer you should get is D.

Sequence of symbols A B C D

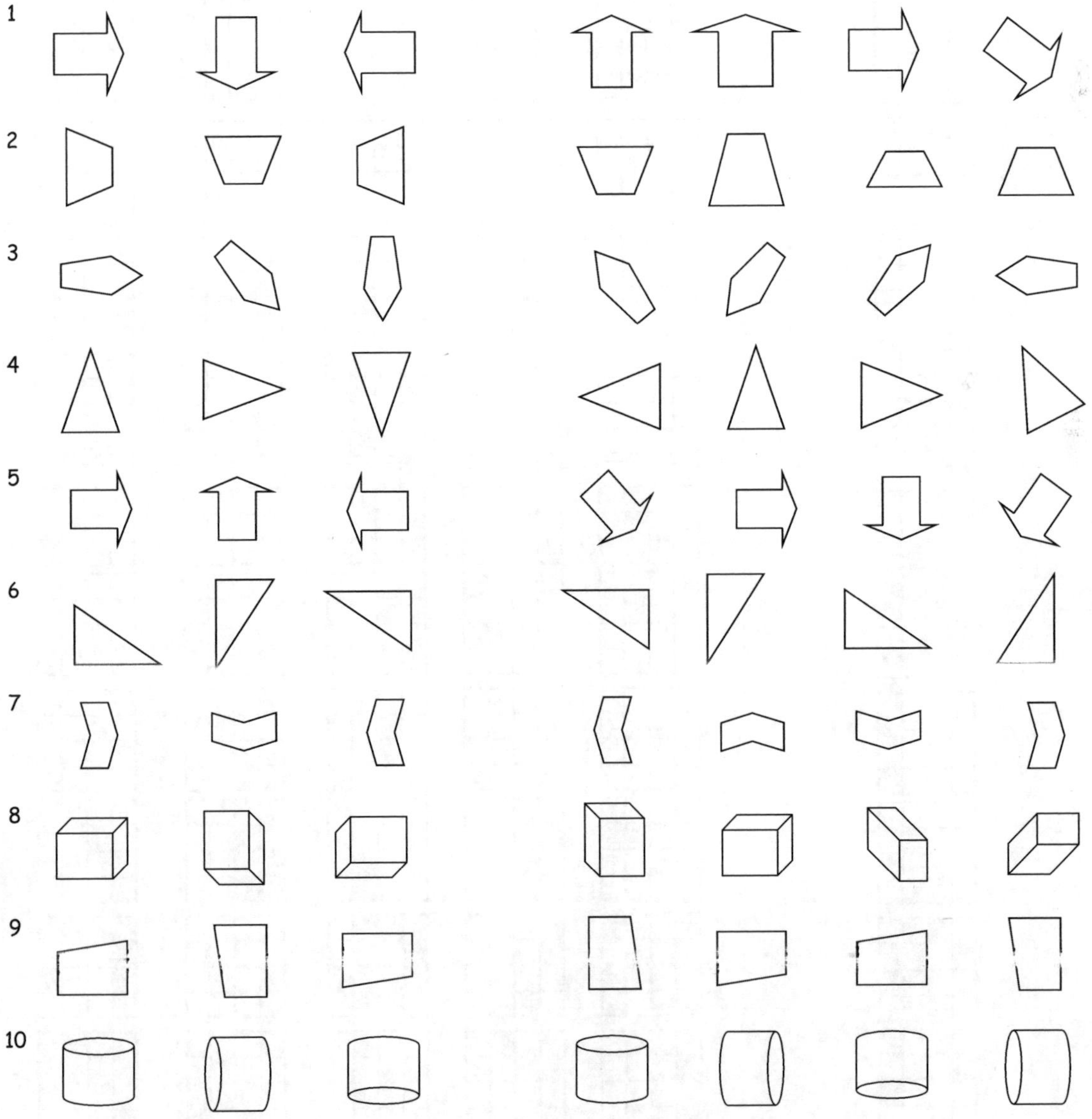

Rotation Worksheet 2. Rotation patterns around a point.

Find the grid from the choices A,B,C,D and E that completes the square grid if the shape is rotated $\frac{1}{2}$ a turn about its centre.

Example: Rotation through $\frac{1}{2}$ a turn. A B C D E

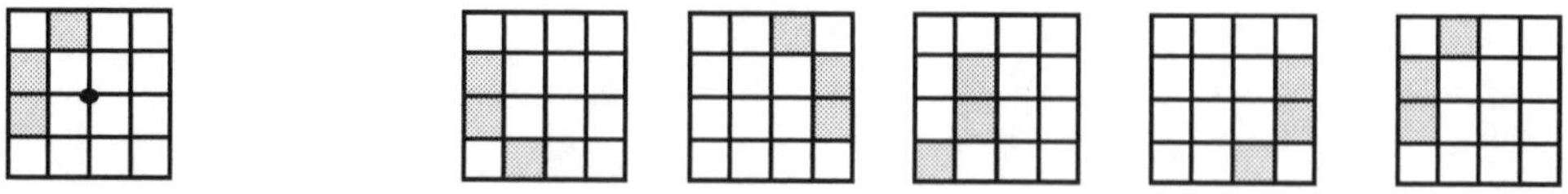

Use tracing paper and rotate the grid to see where the shaded squares move to. The answer you should get is D.

Rotation through $\frac{1}{2}$ a turn. A B C D E

1
2
3
4
5
6
7
8
9
10

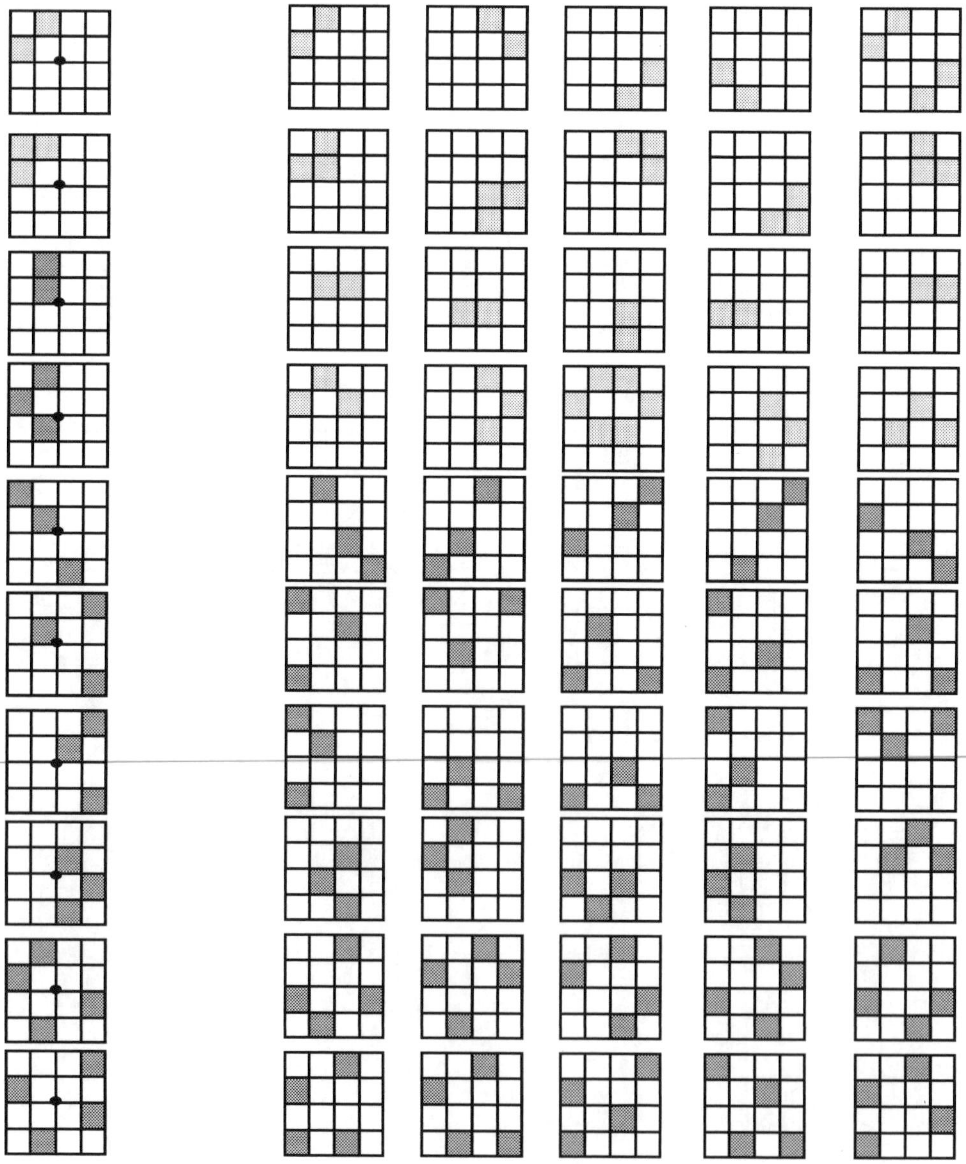

Rotation Worksheet 3. Rotation patterns around a point.

Find the grid from the choices A,B,C,D and E that completes the square grid if the shape is rotated $\frac{1}{4}$ a turn about its centre in a clockwise direction.

Example: Rotation through $\frac{1}{4}$ a turn. A B C D E

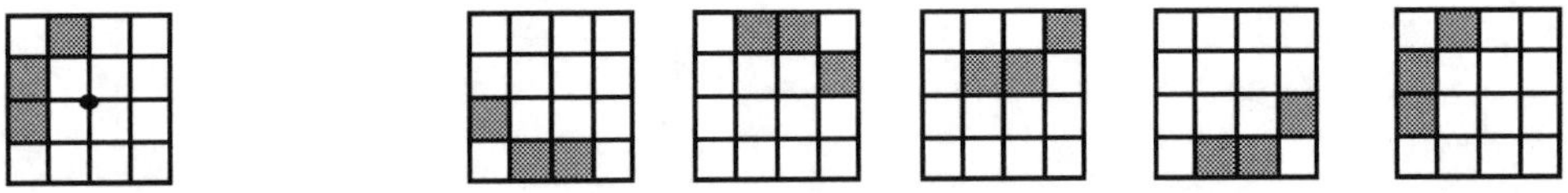

Using tracing paper and the point of your pencil on the centre of the grid and rotating clockwise to find where the squares rotate to. The answer you should get is B.

Now try these.... A B C D E

1
2
3
4
5
6
7
8
9
10

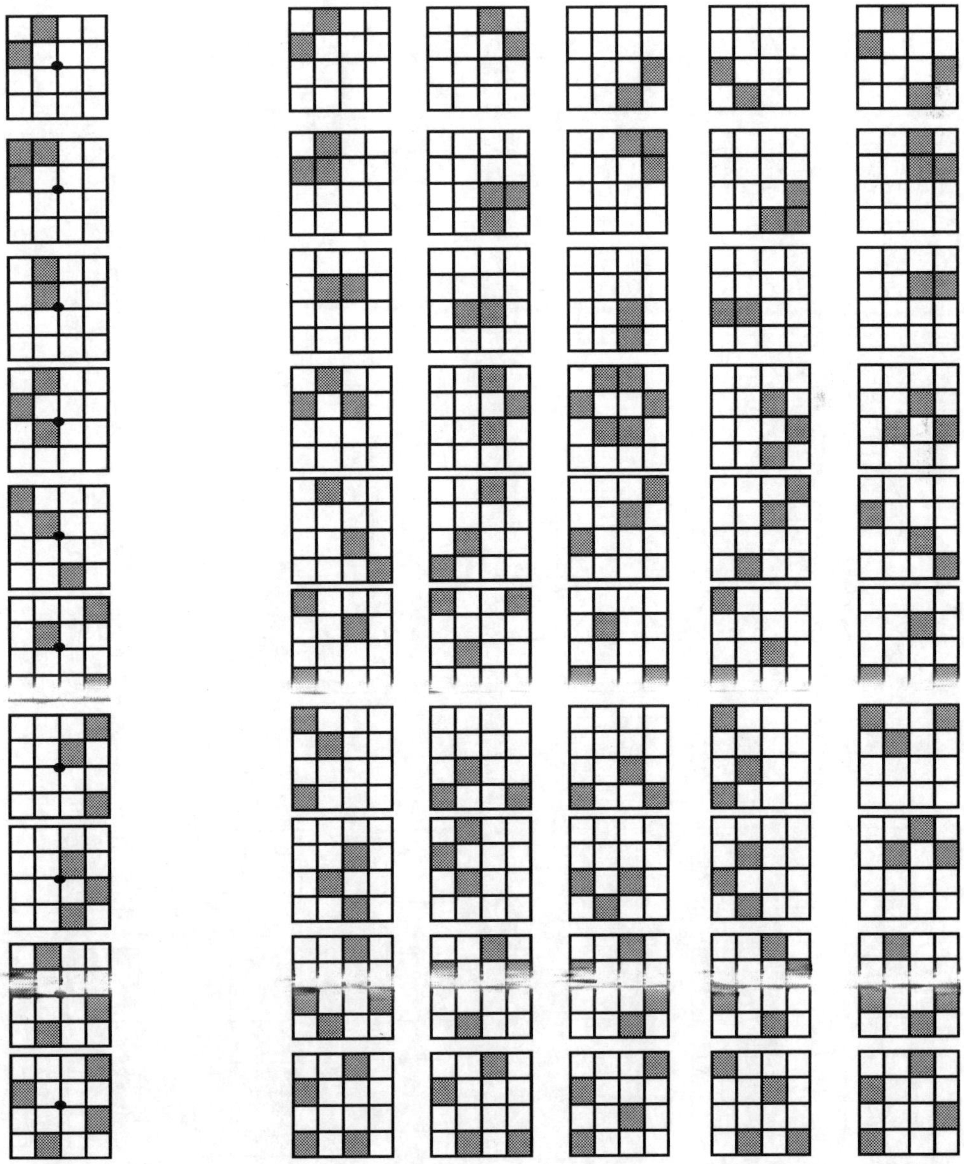

Rotation Worksheet 4. Shape rotation and position.

Here the first shape has been rotated in some way into the second shape. Another shape follows the same rotation to form one of the shapes A,B,C or D.

Example: How are they rotated? A B C D

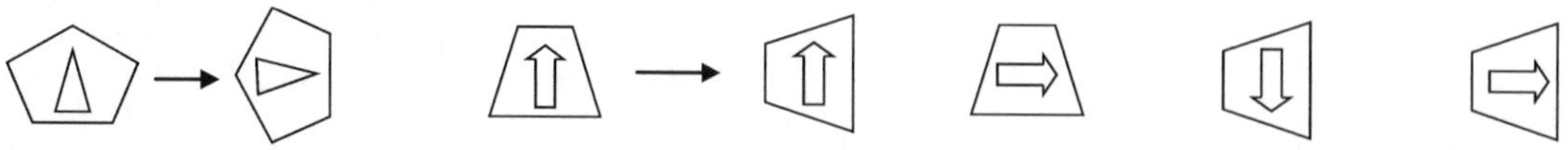

The pentagon is rotated through a $\frac{1}{4}$ turn anticlockwise whilst the triangle is rotated a $\frac{1}{4}$ turn clockwise. Follow the same steps for the trapezium and the arrow. You should get shape B.

How are they rotated? A B C D

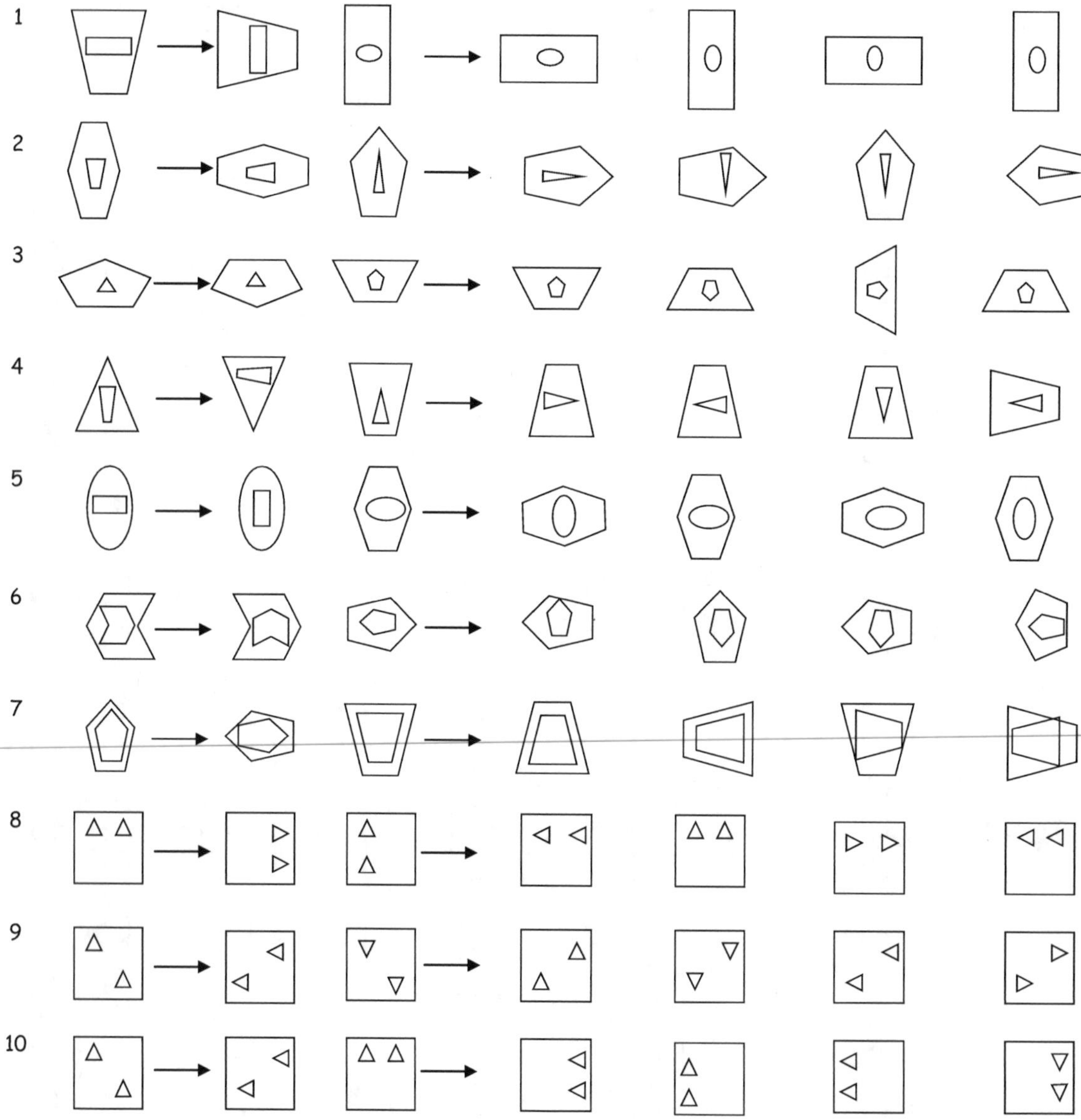

Section 4. Similarity and Congruency.

Example 1. Finding similar shapes.

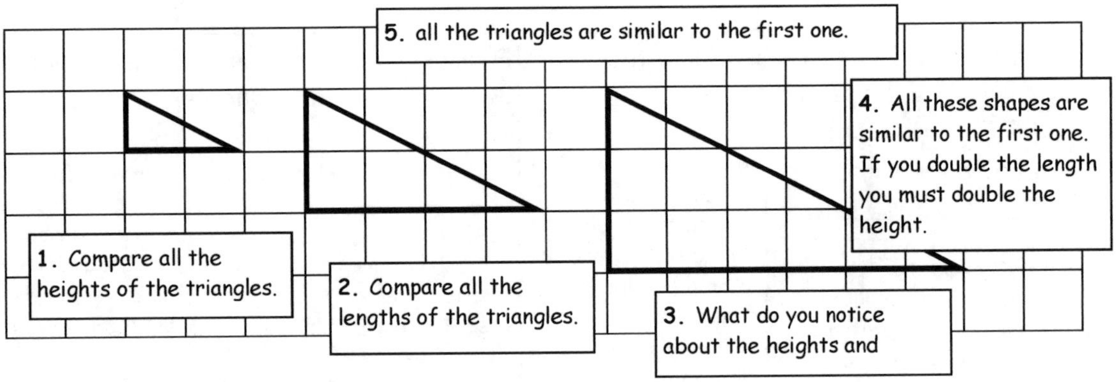

5. all the triangles are similar to the first one.
4. All these shapes are similar to the first one. If you double the length you must double the height.
1. Compare all the heights of the triangles.
2. Compare all the lengths of the triangles.
3. What do you notice about the heights and

Here are the triangles again.

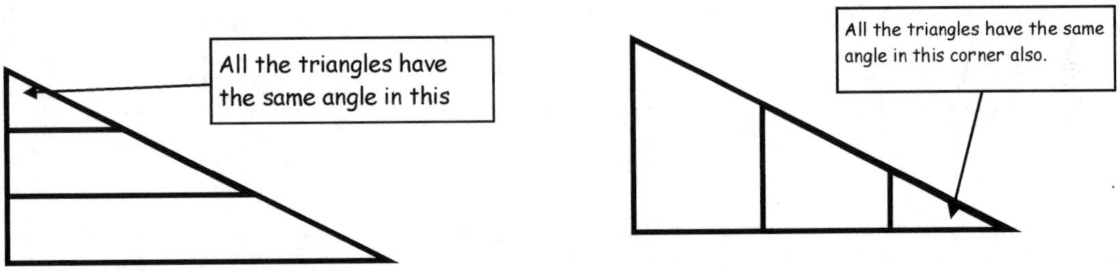

All the triangles have the same angle in this
All the triangles have the same angle in this corner also.

Example 2. Finding Congruency or congruent shapes.

Have a look at the triangles below. Which triangles are congruent to the first one.

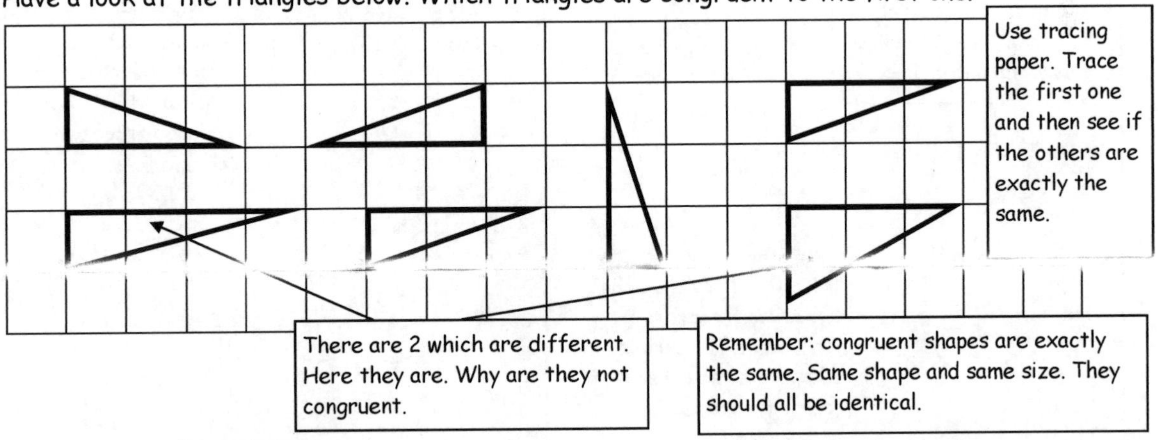

Use tracing paper. Trace the first one and then see if the others are exactly the same.
There are 2 which are different. Here they are. Why are they not congruent.
Remember: congruent shapes are exactly the same. Same shape and same size. They should all be identical.

Similar shapes are enlargements or reductions of the same shape.
Congruent shapes are exactly the same size and shape
How can you tell that these triangles are

How does similarity and congruence help to improve non verbal reasoning?

Here are the Straight A Students who are going to show you what to look for when solving non verbal reasoning problems.

Example 1. See if you agree with how the first shape goes into the second shape.

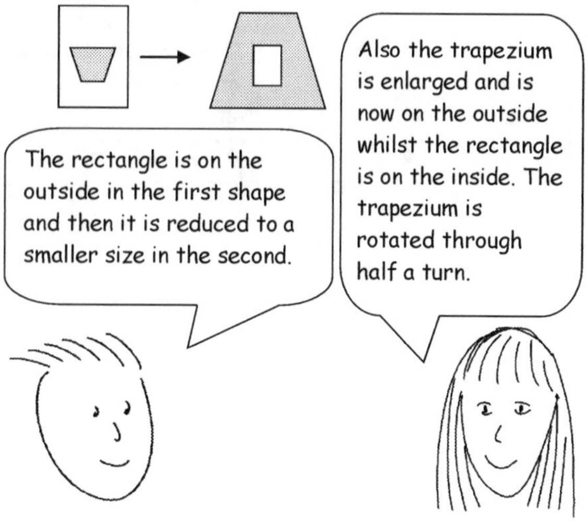
The rectangle is on the outside in the first shape and then it is reduced to a smaller size in the second.
Also the trapezium is enlarged and is now on the outside whilst the rectangle is on the inside. The trapezium is rotated through half a turn.

Example 2. Here are three shapes. What could be the next in the sequence?

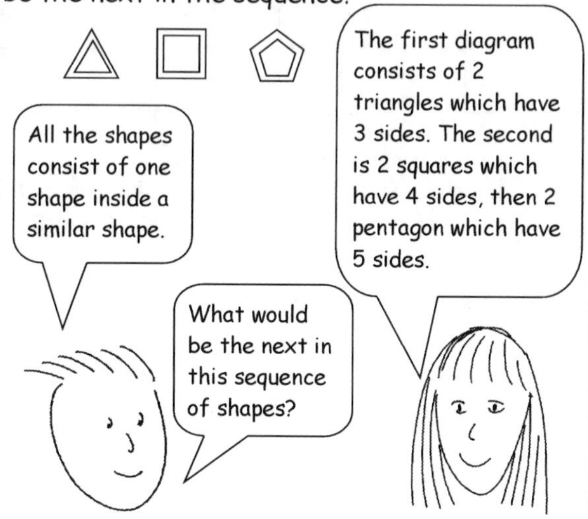
All the shapes consist of one shape inside a similar shape.
The first diagram consists of 2 triangles which have 3 sides. The second is 2 squares which have 4 sides, then 2 pentagon which have 5 sides.
What would be the next in this sequence of shapes?

Example 3. How does the first shape transform into the second? Here is how it was explained.

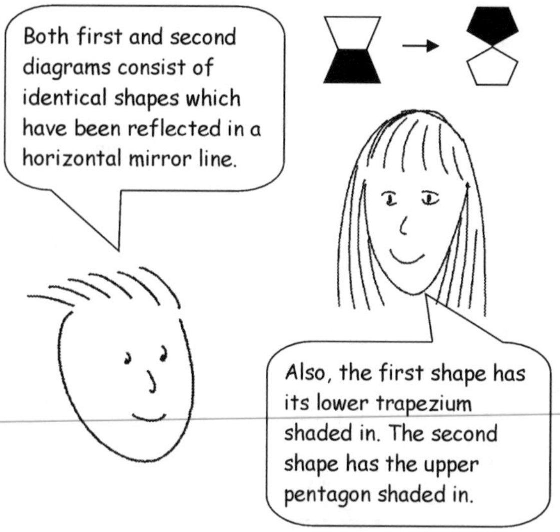
Both first and second diagrams consist of identical shapes which have been reflected in a horizontal mirror line.
Also, the first shape has its lower trapezium shaded in. The second shape has the upper pentagon shaded in.

Example 4. Can you explain how the first shape has been transformed into the second?

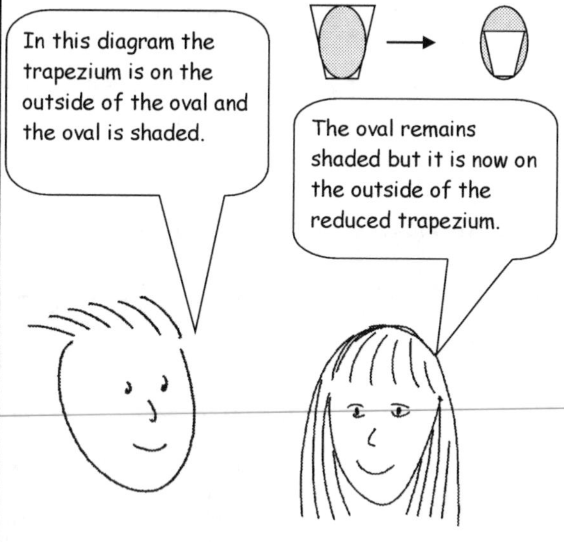
In this diagram the trapezium is on the outside of the oval and the oval is shaded.
The oval remains shaded but it is now on the outside of the reduced trapezium.

Here is one for you to try.
Write down in the space below how the first diagram changes into the second?
Which shapes stay the same?
Which shapes change size?

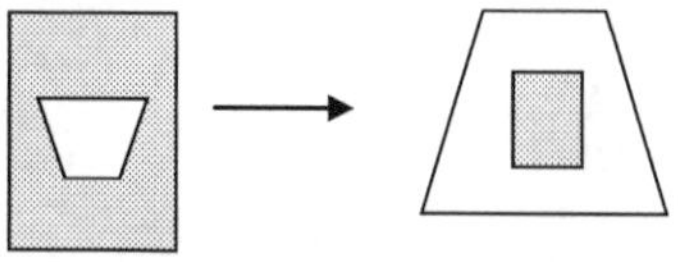

....

....

Similarity Practice 1.

Circle the shapes which are similar to the first one.

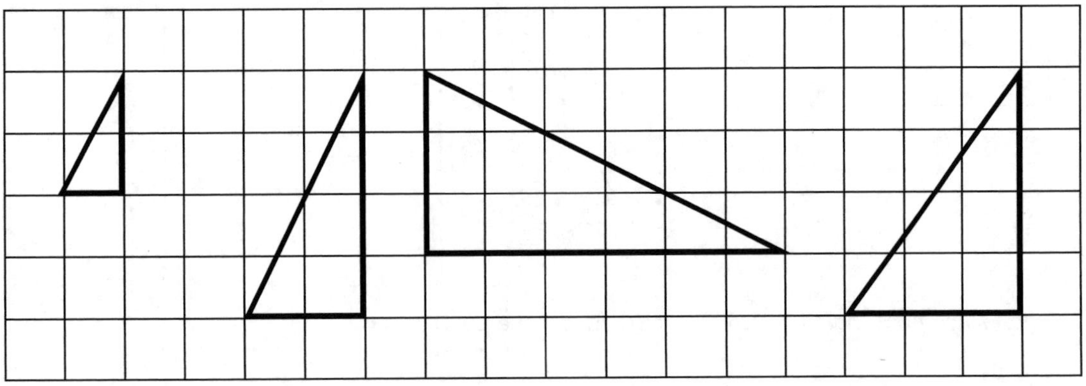

Circle the shapes which are congruent.

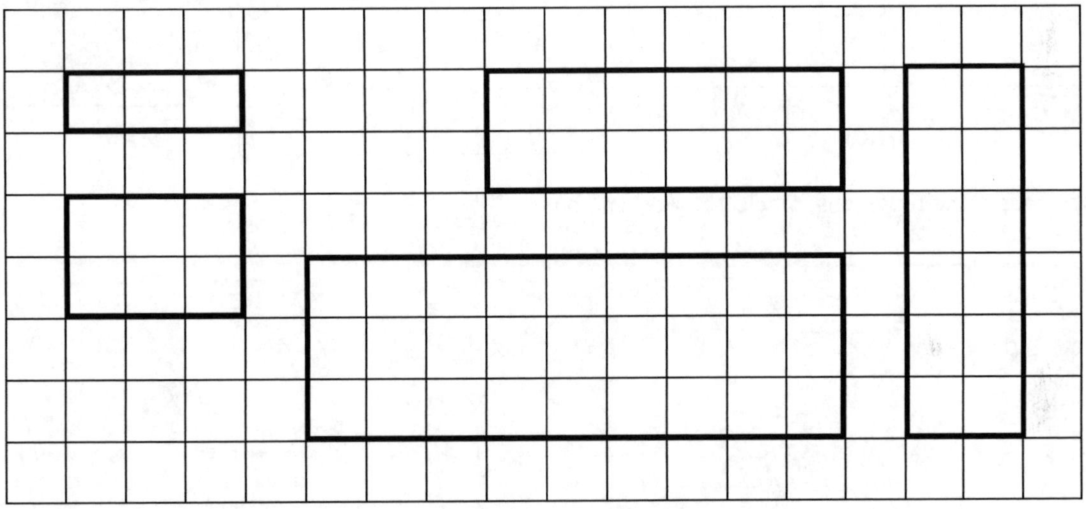

Enlarge each of these by making them twice as large.

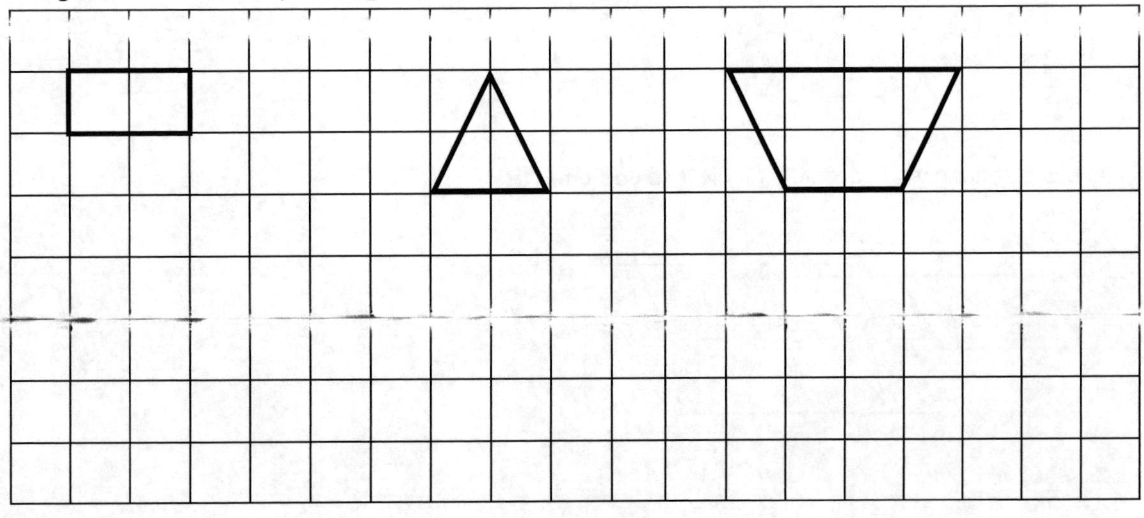

Similarity Practice 2.

Here are some triangles. Circle the odd shape out.

Here are some rectangles. Circle the odd shape one out.

Here are some hexagons. Circle the odd one out.

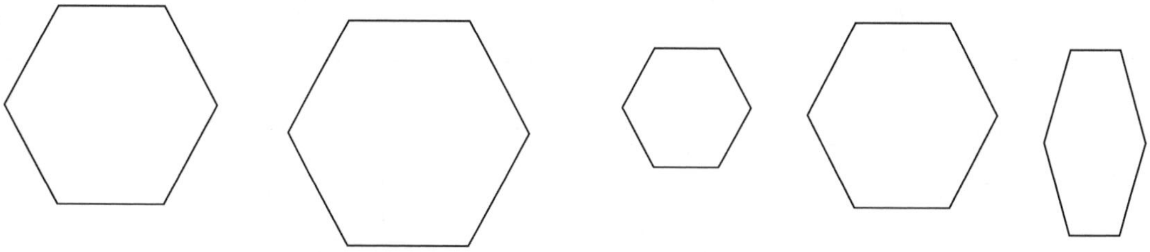

Here are some trapeziums. Circle the odd one out.

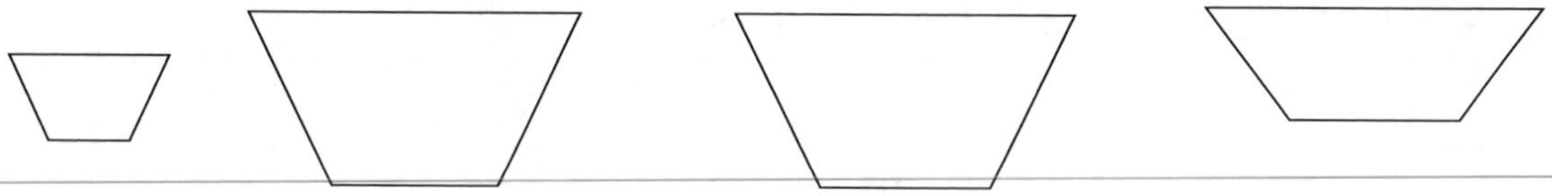

Here are some parallelograms. Circle the odd one out.

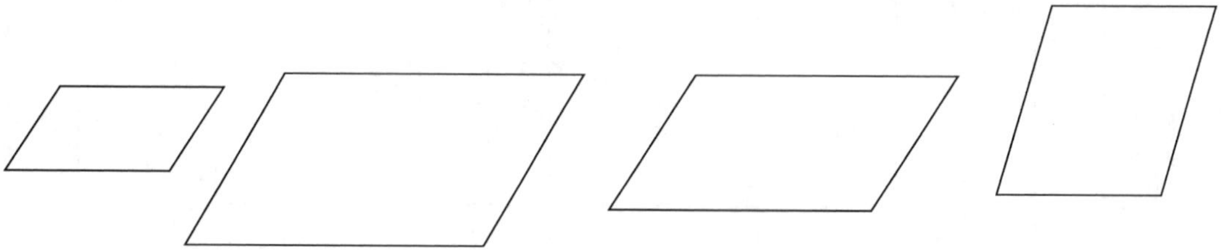

Similarity and Congruency Worksheet 1.

Find the odd one out in the following sets of shapes. You are looking for a shape that does not match the others. The odd one out is the one which is not congruent to the others i.e. not exactly the same when it is rotated into the same position as the others.

Example: Find the odd one out? A B C D E

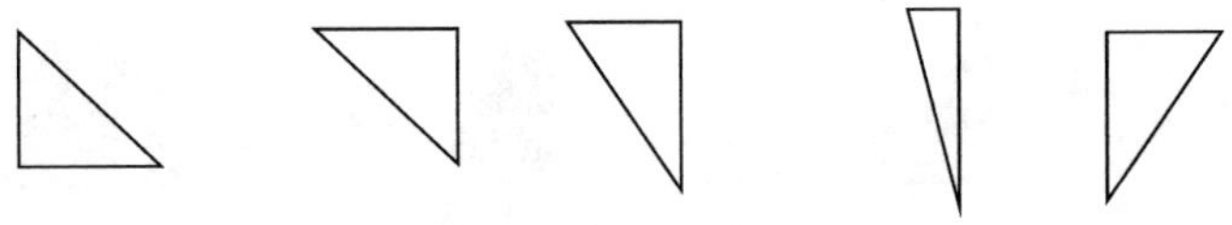

You should see that shape D is different from the others. It is not congruent to the other triangles.

Find the odd one out?

A B C D E

1

2

3

4

5

6

7

8

9

10

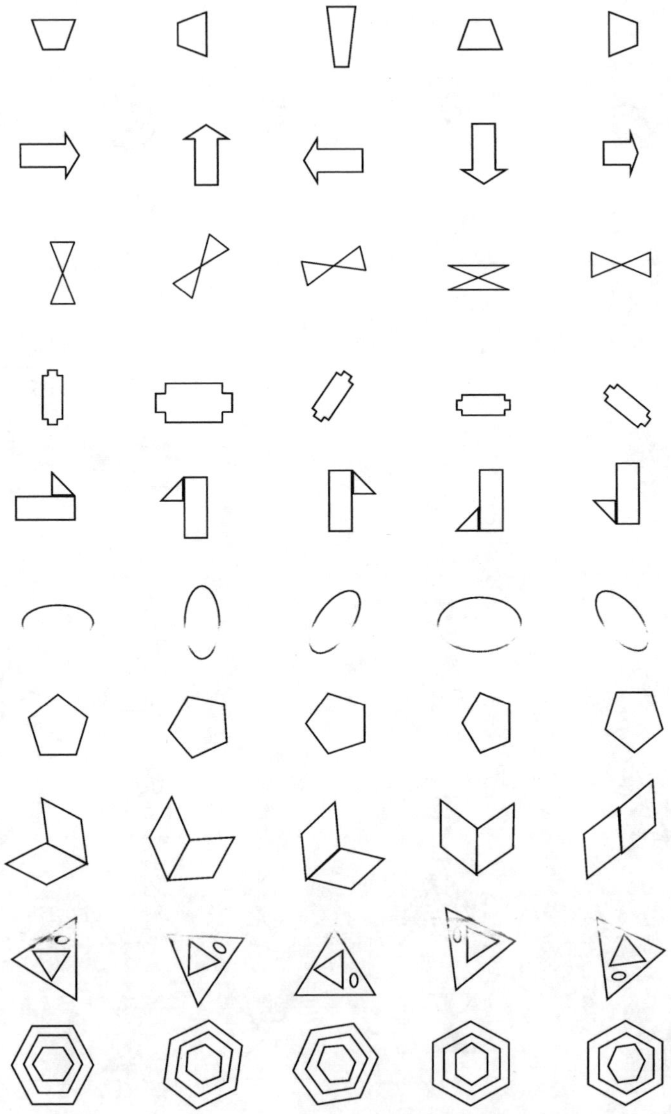

Similarity and Congruency Worksheet 2.

The first shape is transformed into a similar shape. Find the shape labelled A,B,C or D that goes with the first two shapes. You need to decide how they are the same and then choose which one fits best.

Example: How does it change? A B C D

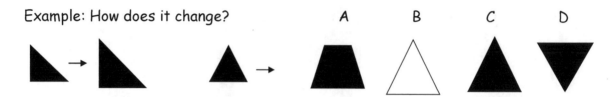

The first shape is a right angled triangle and it is enlarged and stays shaded in. The next shape is an equilateral triangle and so following the same change it will be enlarged into shape C. So the answer you should get is C.

How does it change? A B C D

1

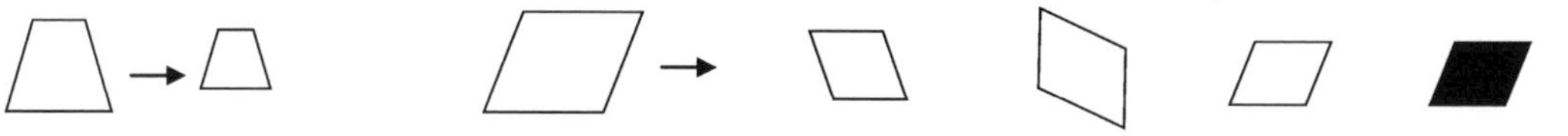

2

3

4

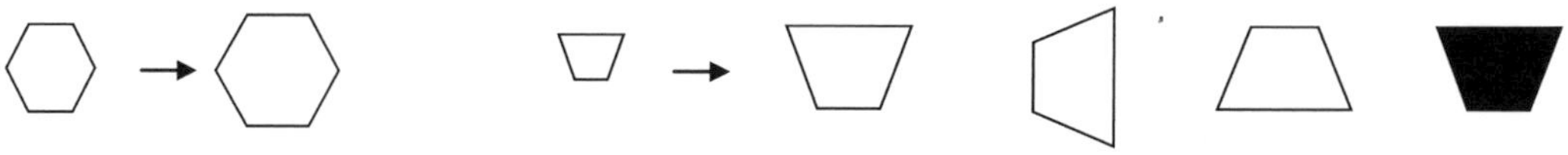

5

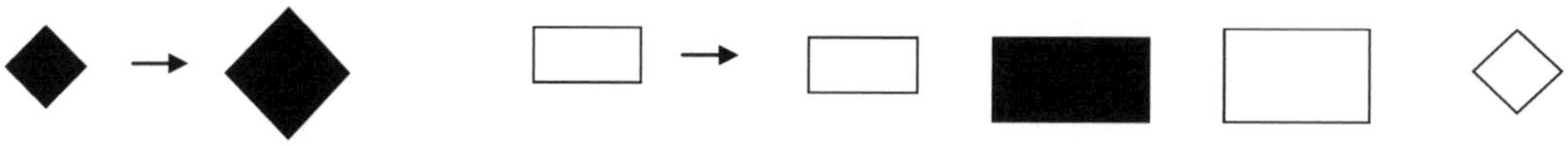

6

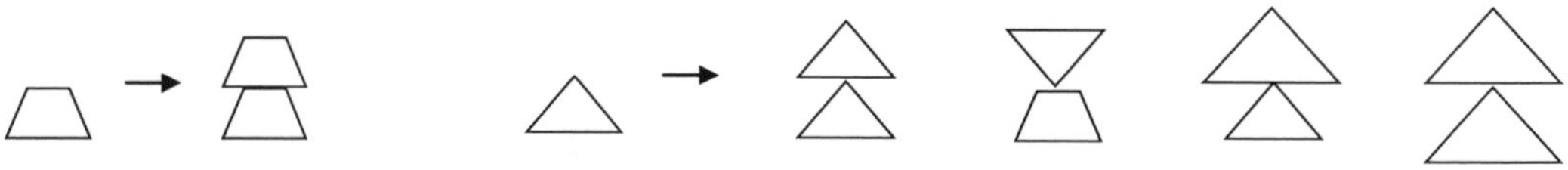

7

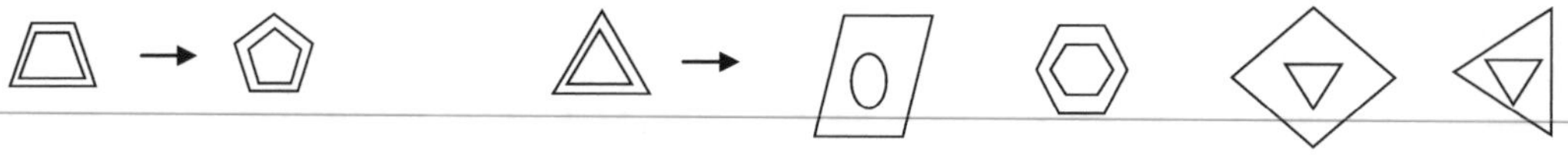

8

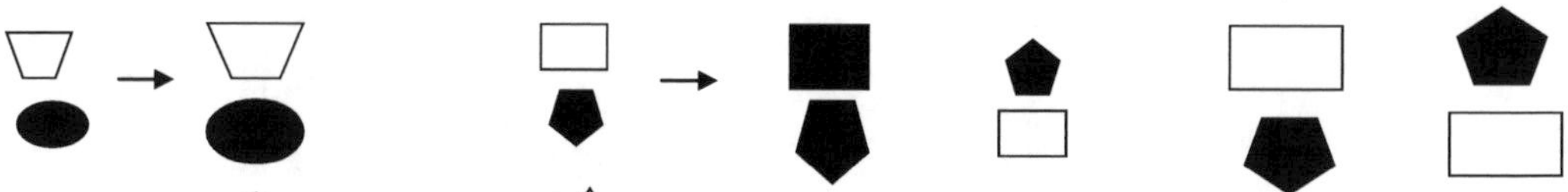

9

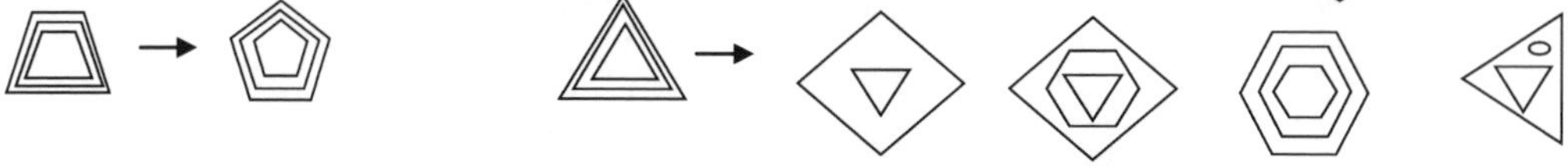

10

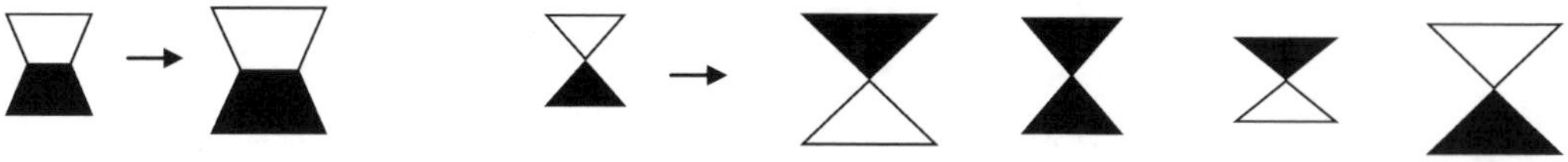

Section 5. Mixed Non Verbal Reasoning.

Here are the Straight A Students who are going to explain some mixed examples and point out what to look for when solving different types of non verbal reasoning problems.

Example 1. Look at the following sequence of diagrams and see if you can anticipate what could be a continuation of a sequence of shape.

Each of the shapes consists of 2 shapes one above the other, which themselves are different to each other.
In each pair of shapes the top shape is shaded and the bottom shape is un-shaded. What could come next?

Example 2. Look at the grids to the right here. Can you explain how the pattern will continue?

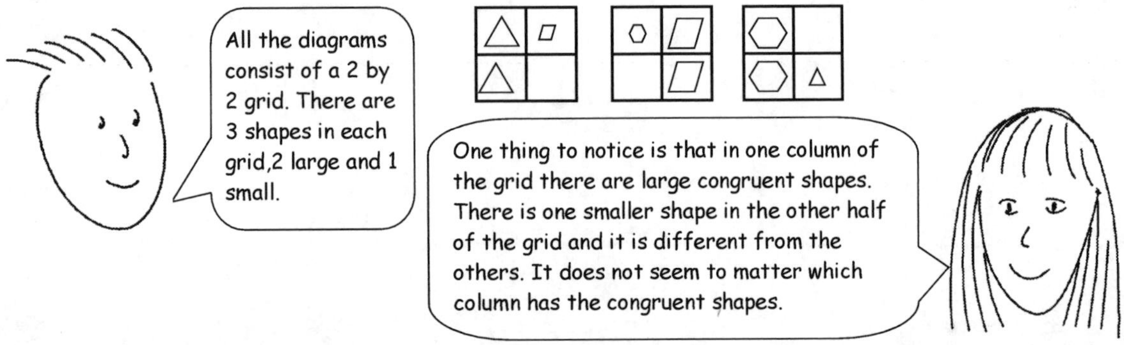
All the diagrams consist of a 2 by 2 grid. There are 3 shapes in each grid,2 large and 1 small.
One thing to notice is that in one column of the grid there are large congruent shapes. There is one smaller shape in the other half of the grid and it is different from the others. It does not seem to matter which column has the congruent shapes.

Example 3. How does the first shape transform into the second? Here is how it is explained.

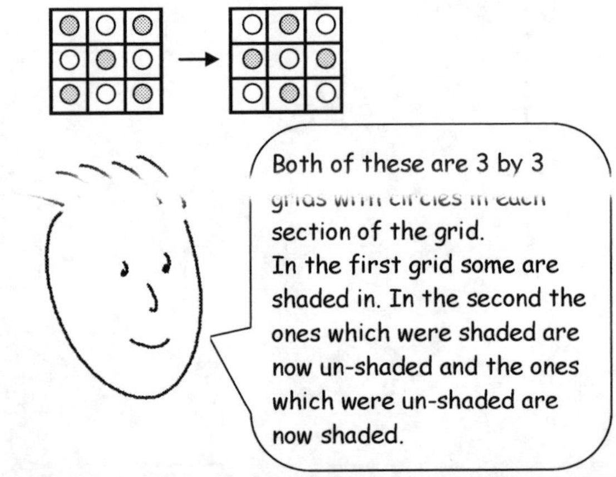
Both of these are 3 by 3
section of the grid.
In the first grid some are shaded in. In the second the ones which were shaded are now un-shaded and the ones which were un-shaded are now shaded.

Example 4. Can you explain how the first shape has been transformed into the second?

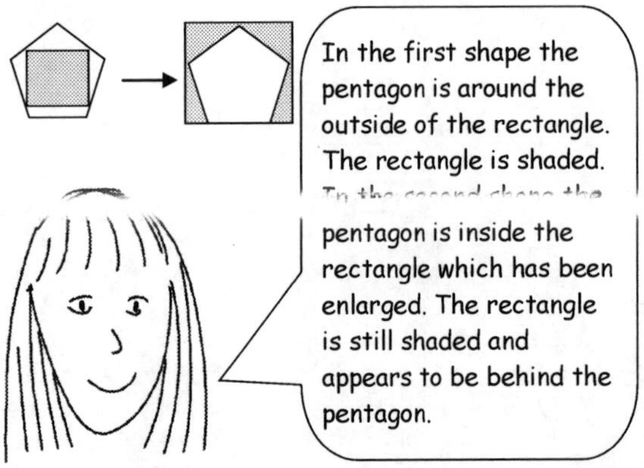
In the first shape the pentagon is around the outside of the rectangle. The rectangle is shaded.
pentagon is inside the rectangle which has been enlarged. The rectangle is still shaded and appears to be behind the pentagon.

Here is one for you to try.

Which shapes stay the same?

Which shapes change size?

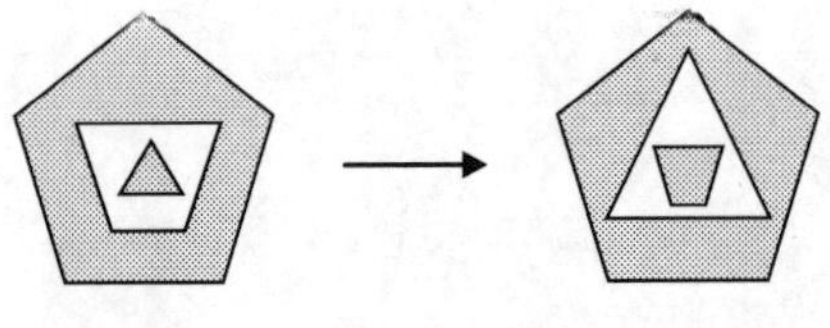

....

....

....

Mixed Non Verbal Worksheet 1.

There are two shapes where the first has been changed into the second. Look at the other shape and find the shape labelled A,B,C or D which would best fit if the shape had been changed in a similar way. You need to decide how they are changed in the example and make the same kind of change and then choose which one fits best.

Example: How does it change? A B C D

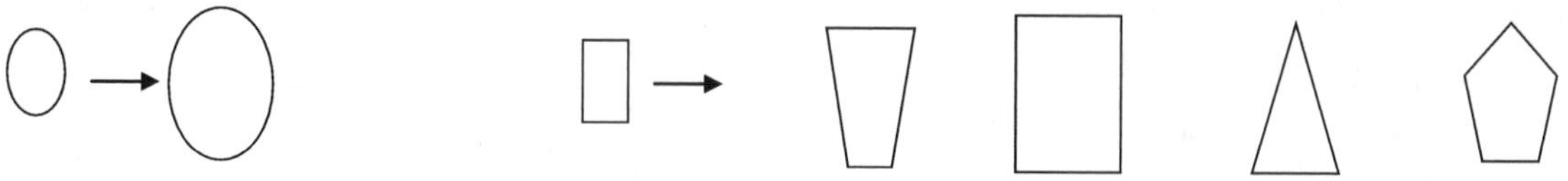

The first oval shape has been enlarged into the second oval. So using the same rule to the rectangle you need to enlarge it and the answer you should get is B.

How does it change? A B C D

1

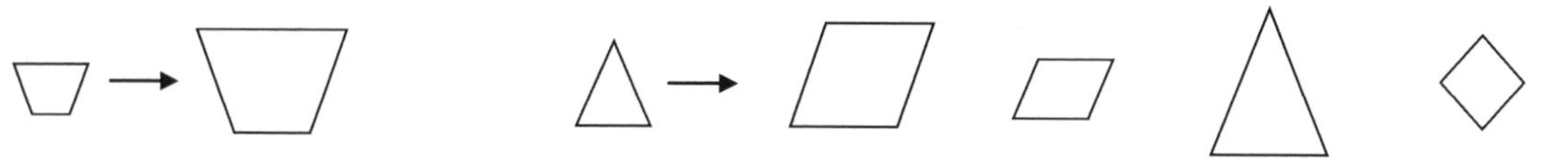

2

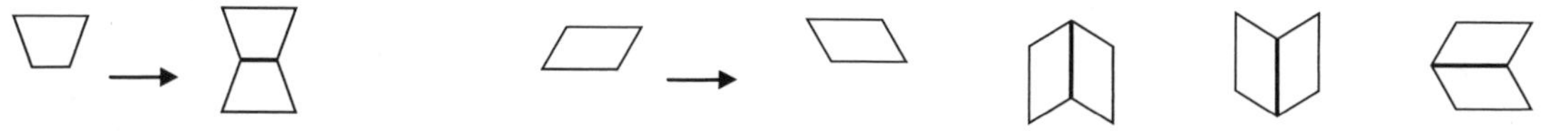

3

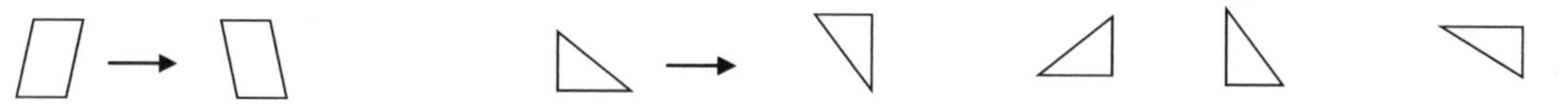

4

5

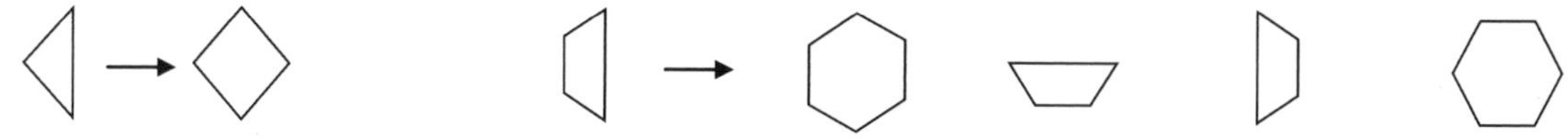

6

7

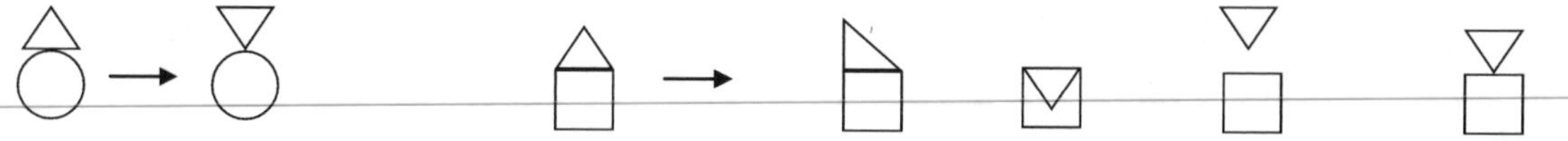

8

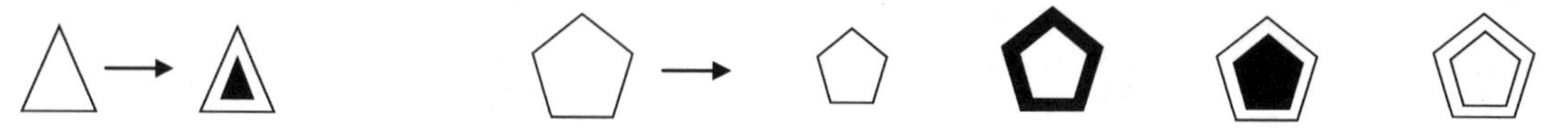

9

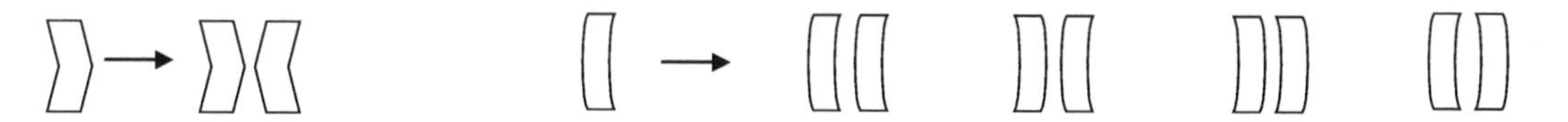

10

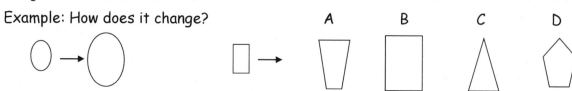

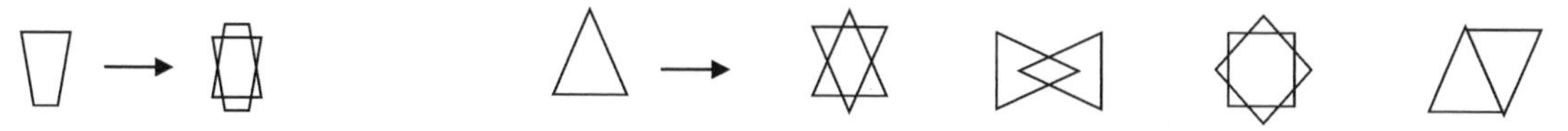

Mixed Non Verbal Worksheet 2.

The first three shapes are similar in one or more ways. Find the shape labelled A,B,C or D that goes with the three shapes. You need to decide how they are the same and then choose which one fits best.

Example: A sequence of symbols.

A B C D

All the shapes in the sequence are triangles which are fully shaded in. Therefore the one you should be looking for is a triangle which is fully shaded in. The answer you should get is A.

A sequence of symbols.

A B C D

1
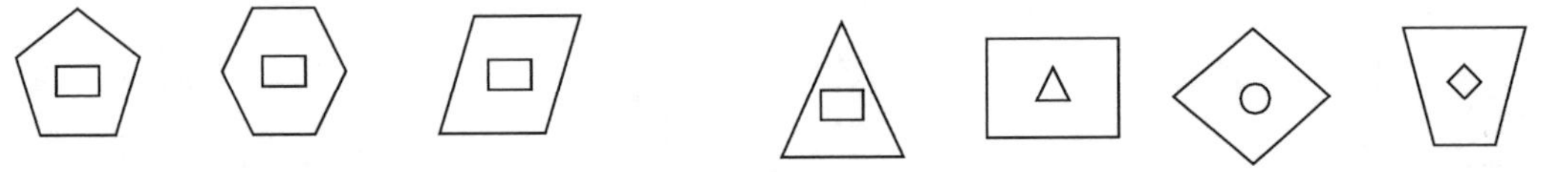

2

3

4
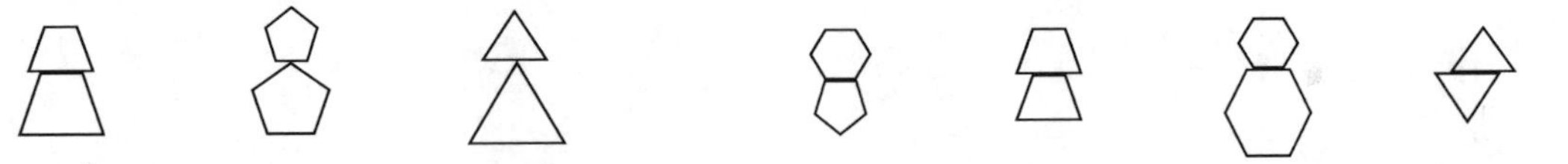

5
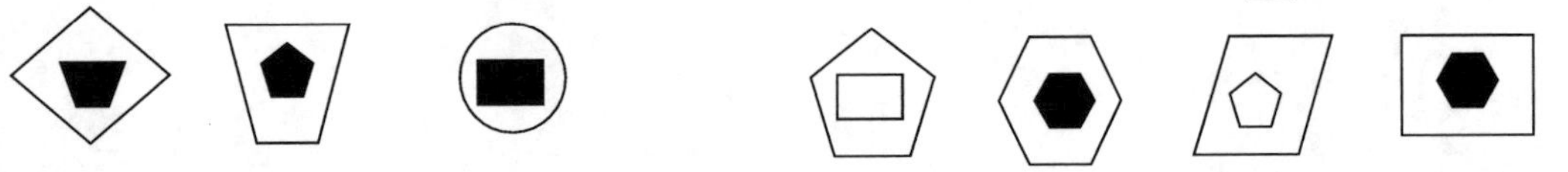

6
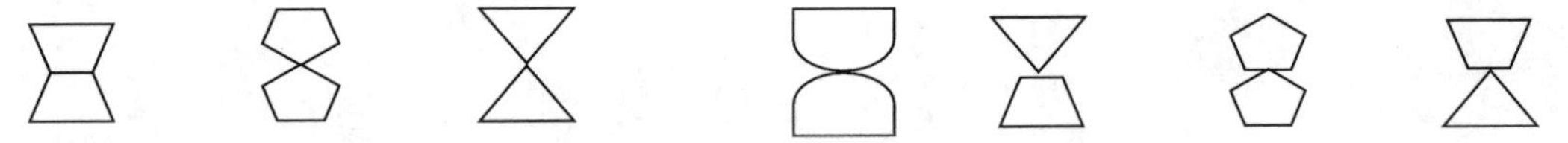

7
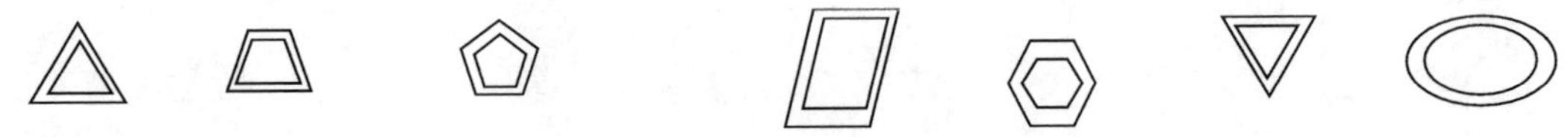

8
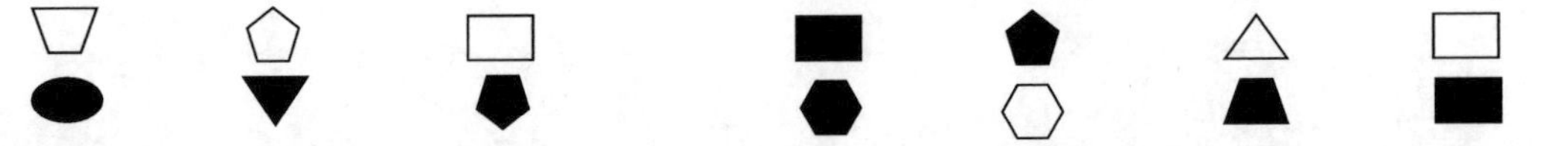

9
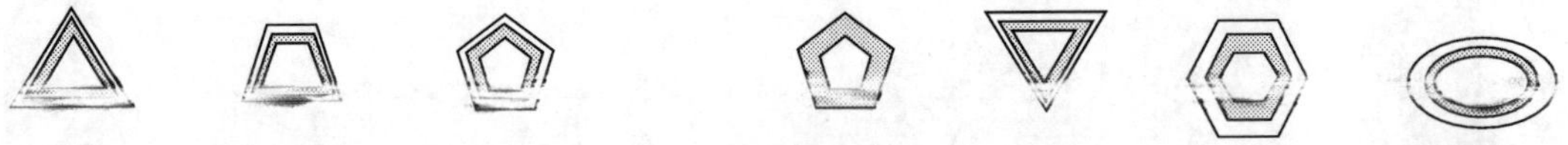

10
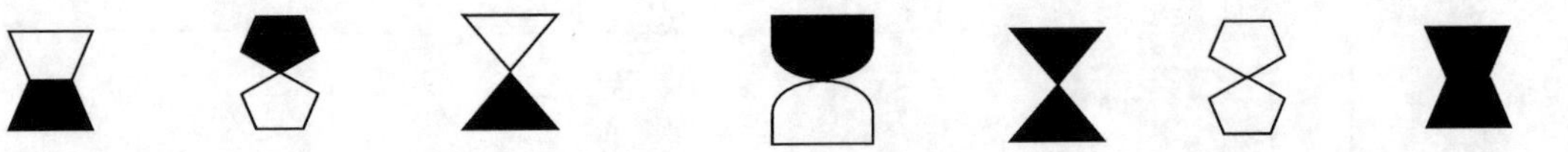

Mixed Non Verbal Worksheet 3.

There are two shapes where the first has been changed into the second. Look at the other shape and find the shape labelled A,B,C or D which would best fit if the shape had been changed in a similar way. You need to decide how they are changed in the example and make the same kind of change and then choose which one fits best.

Example: How does it change? A B C D

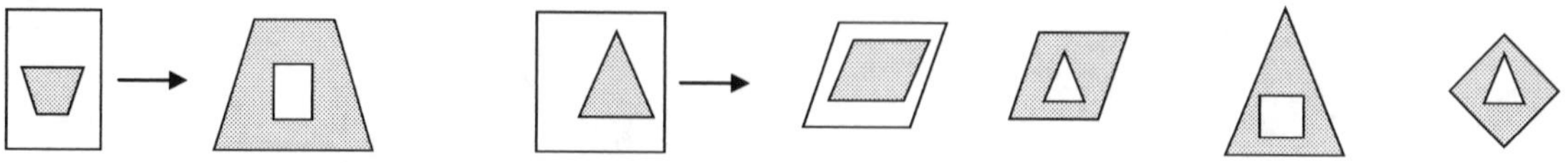

The inside shape is a trapezium and has been enlarged and reflected horizontally or rotated through $\frac{1}{2}$ a turn. The outside is a rectangle and has been reduced in size and has become the inside shape. The trapezium is shaded in the first diagram and stays shaded in the second. The answer you should get is C.

How does it change? A B C D

1 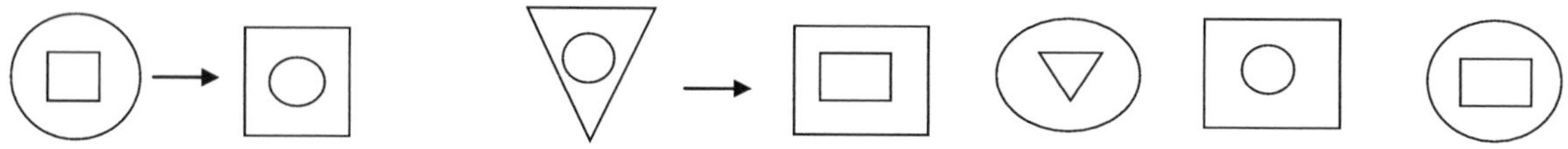

2 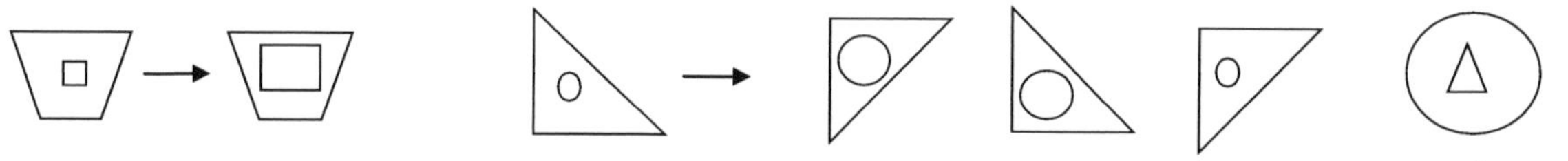

3 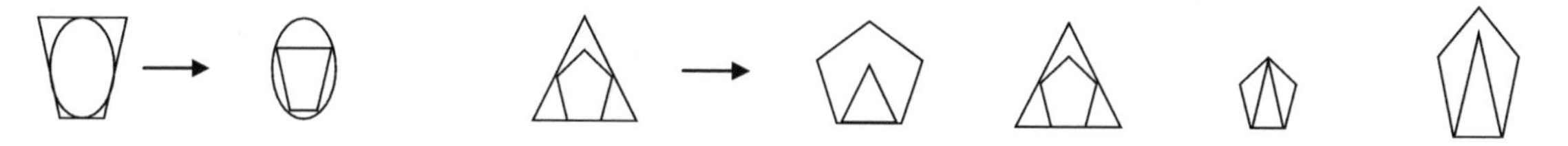

4 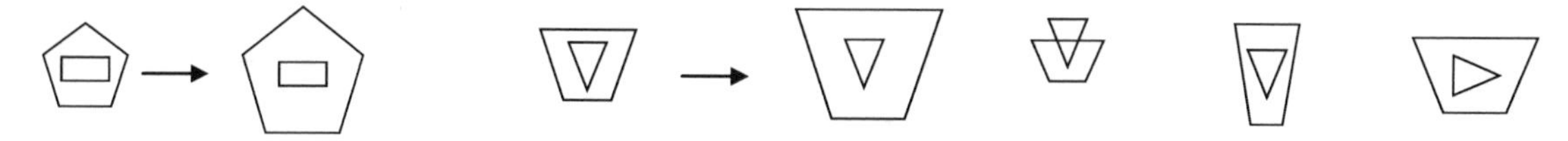

5 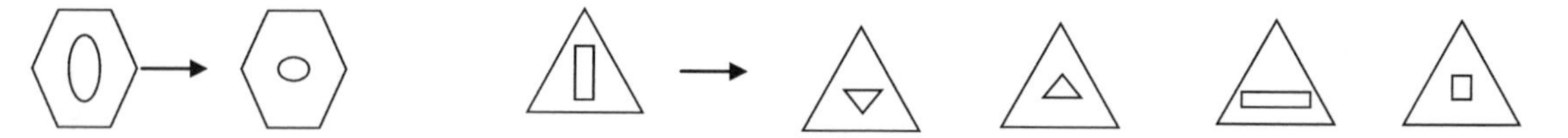

6 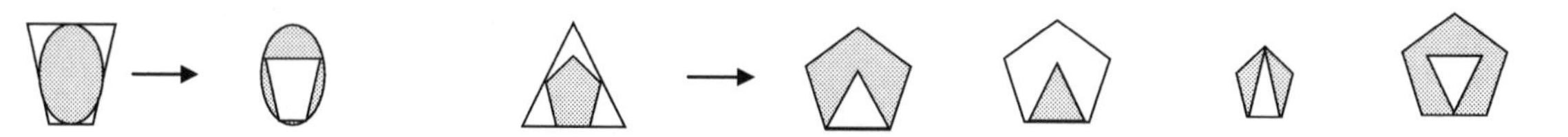

7 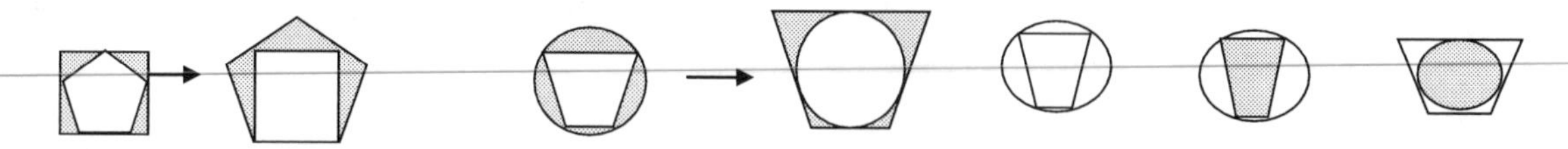

8 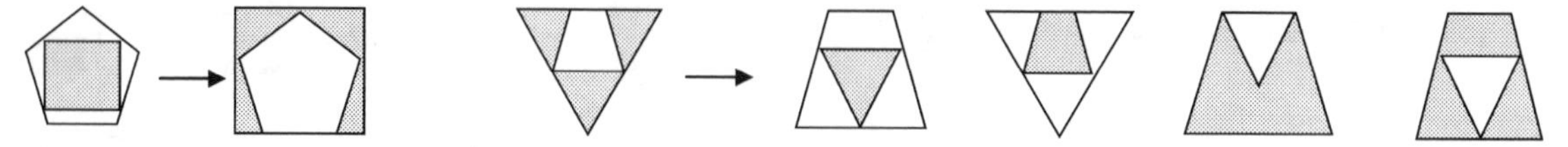

9 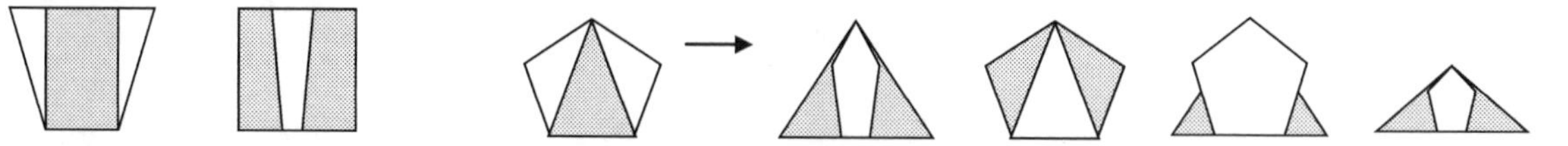

10 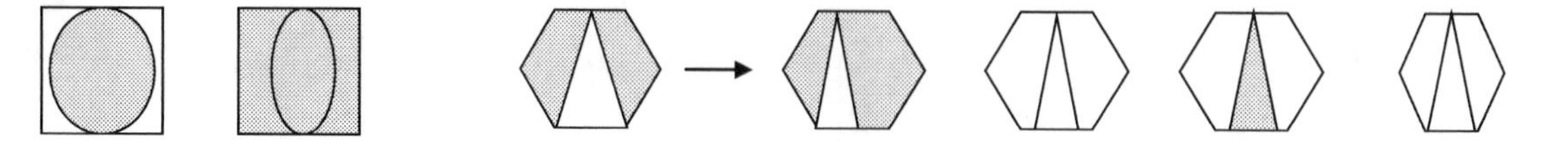

Mixed Non Verbal Worksheet 4.

There are two shapes where the first has been changed into the second. Look at the next shape and find the shape labelled A,B,C or D which would best fit if the shape had been changed in a similar way. You need to decide how they are changed in the example and make the same kind of change and then choose which one fits best.

Example: How does it change?

A B C D

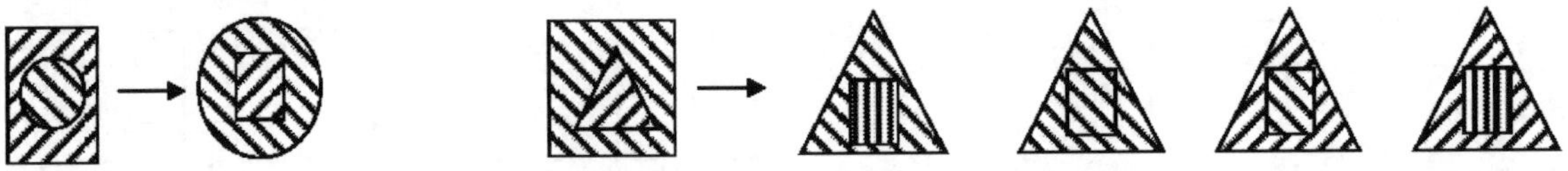

The inside shape is a circle and has been enlarged and becomes the outside shape. The outside is a rectangle and has been reduced in size and has become the inside shape. The first shape has the shading lines rising from left to right in the outside shape falling from left to right in the inside shape. This is reversed in the second shape. So if you apply the same to the new shape the answer you should get is C.

How does it change?

A B C D

1

2
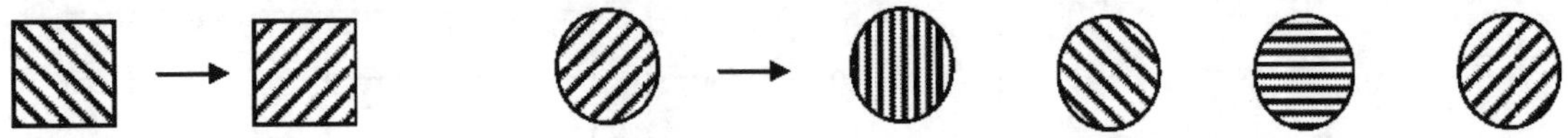

3

4
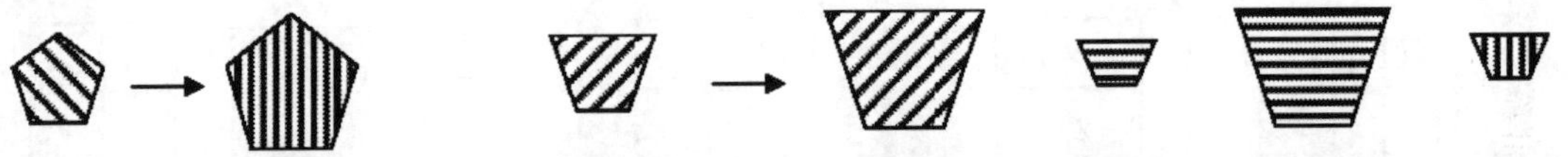

5
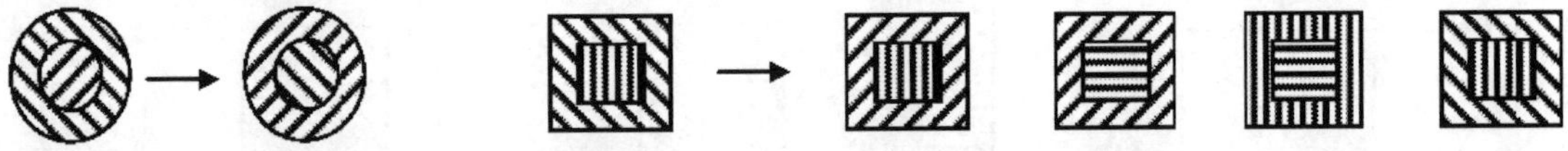

6
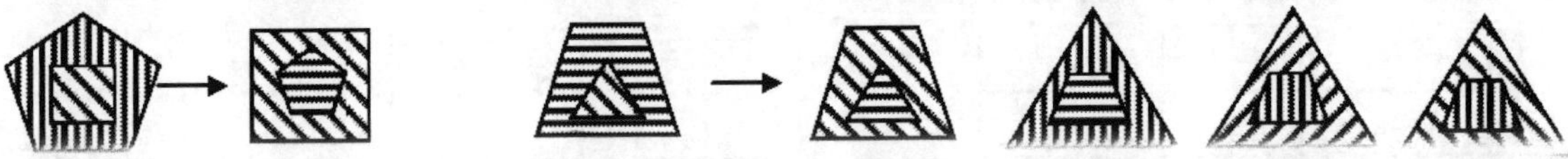

7
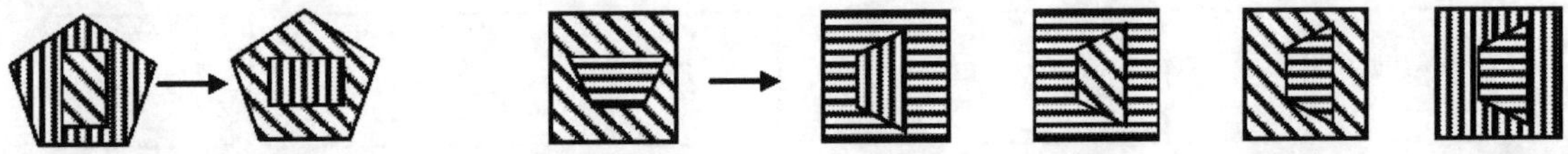

8
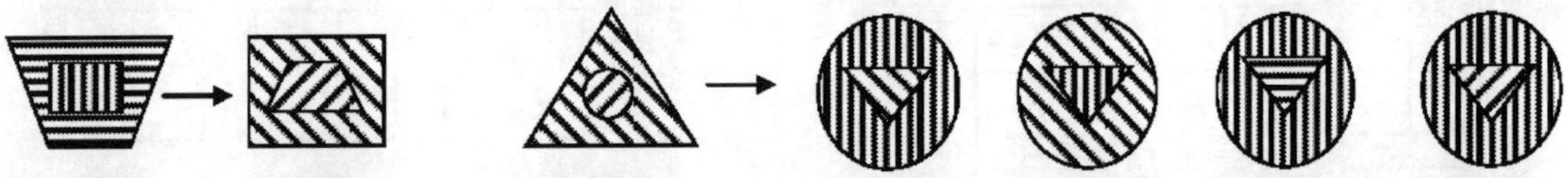

9
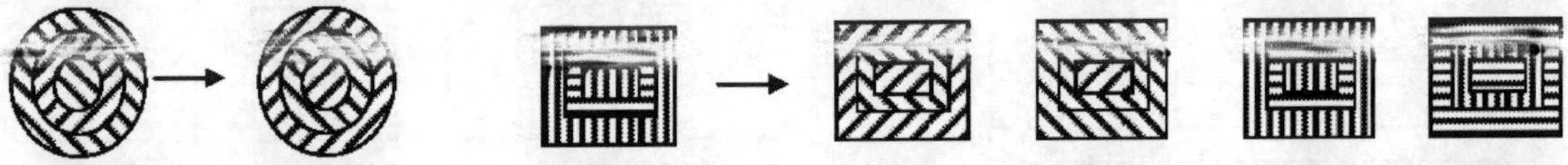

10
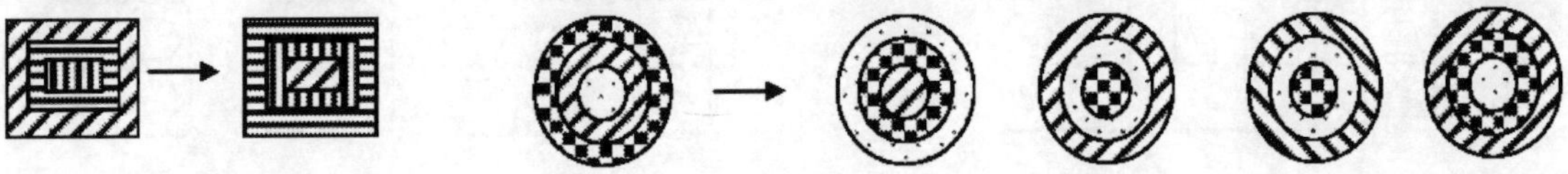

Mixed Non Verbal Worksheet 5.

The first three rectangles or grids are partly fully filled with shapes which are similar in one or more ways. Find the shape labelled A,B,C or D that goes with the three shapes. You need to decide how they are the same and then choose which one fits best.

Example: A sequence of symbols

A B C D

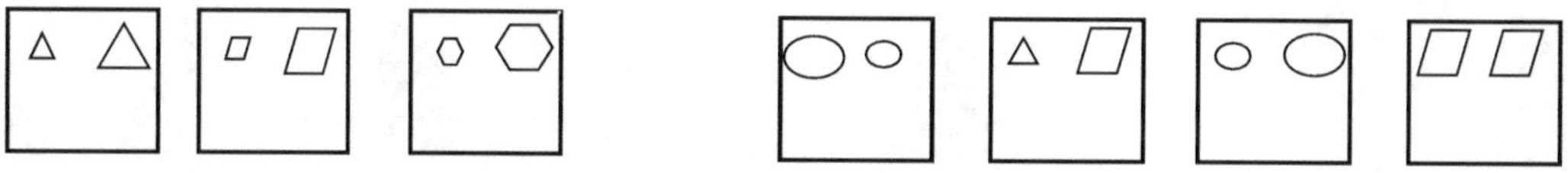

Each grid has a shape in the top left and an enlarged shape of the same kind in the top right. So the only one that fits is the small oval shape and its enlarged oval in the top right. The answer you should get is C.

A sequence of symbols

A B C D

1
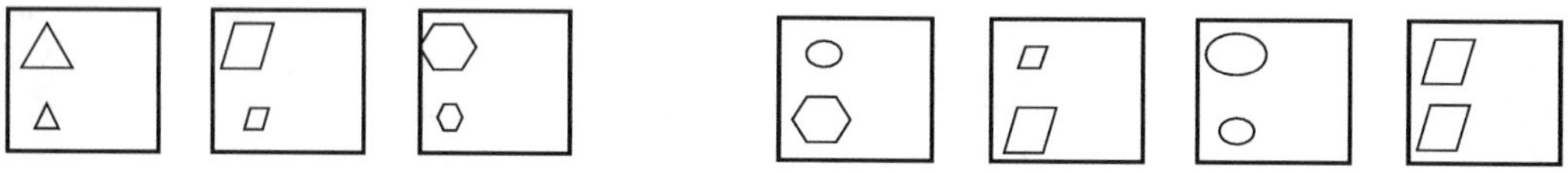

2
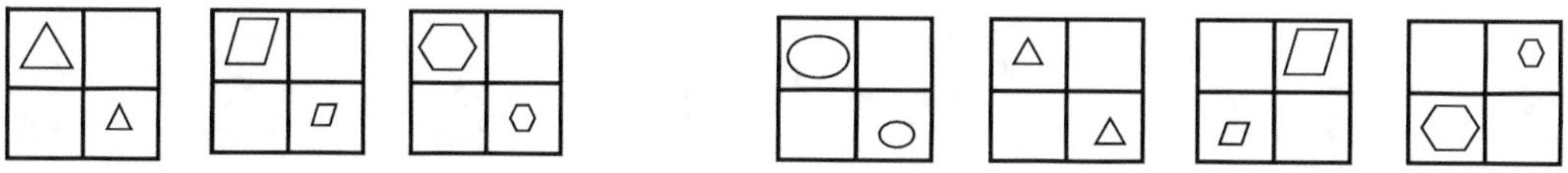

3
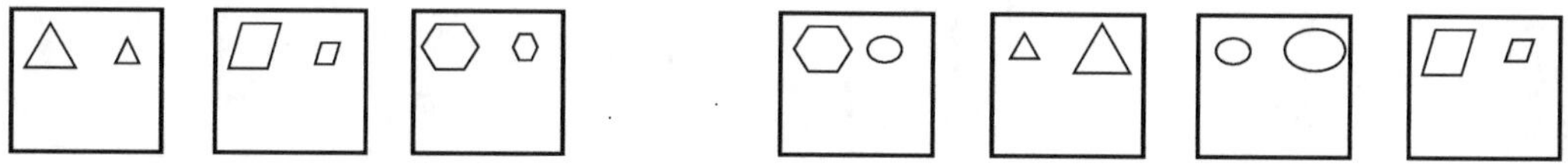

4
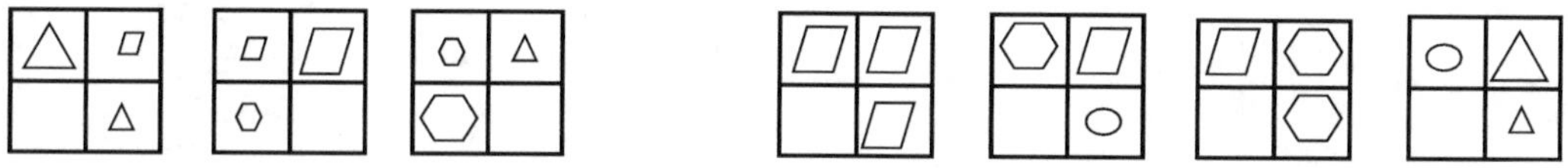

5
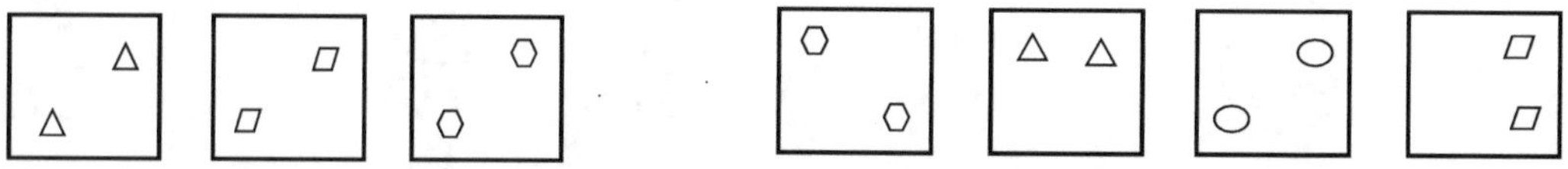

6
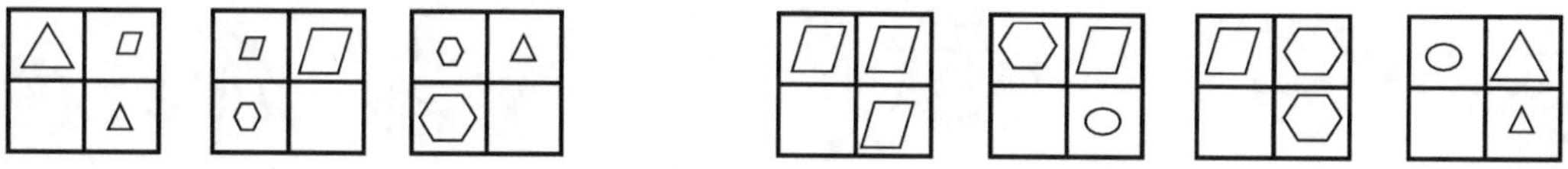

7
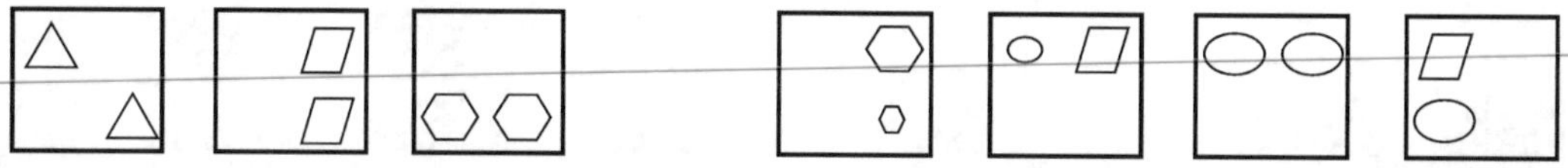

8
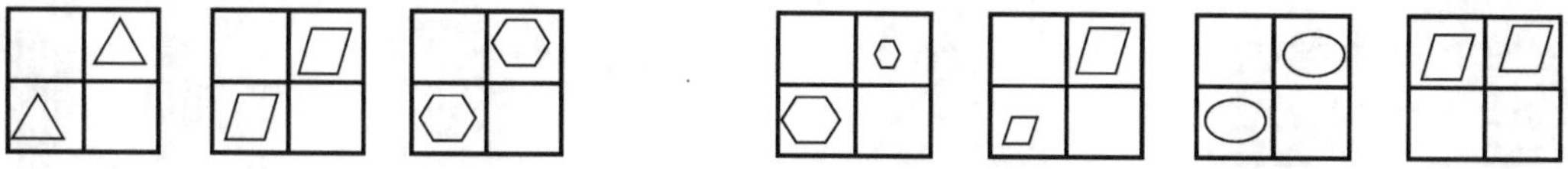

9
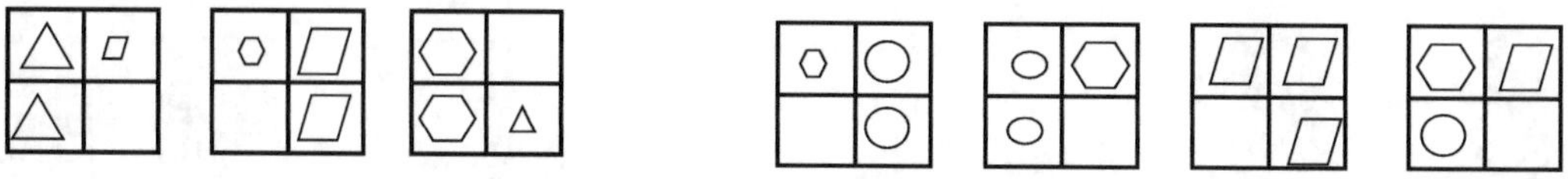

10
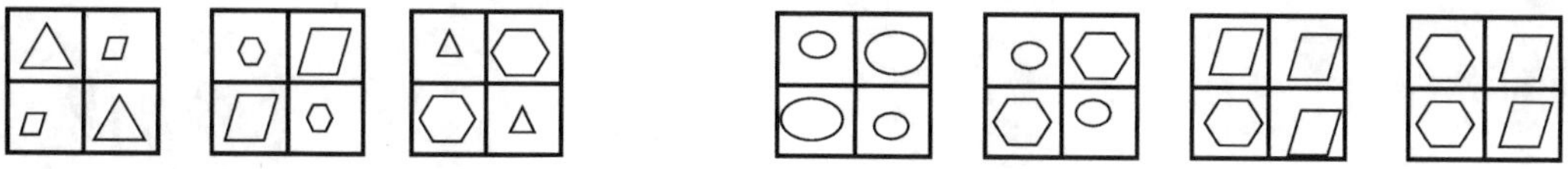

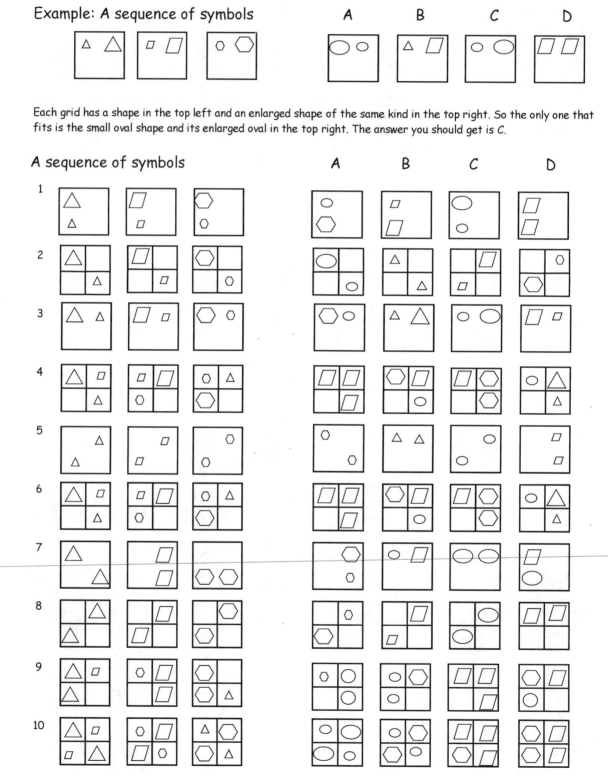

Mixed Non Verbal Worksheet 6.

There are two patterns in a grid where the first has been changed into the second. Look at the new grid and find the grid labelled A,B,C or D which would best fit if the shape had been changed in a similar way. Remember to use the same kind of change and then choose which one fits best.

Example: How does it change? A B C D

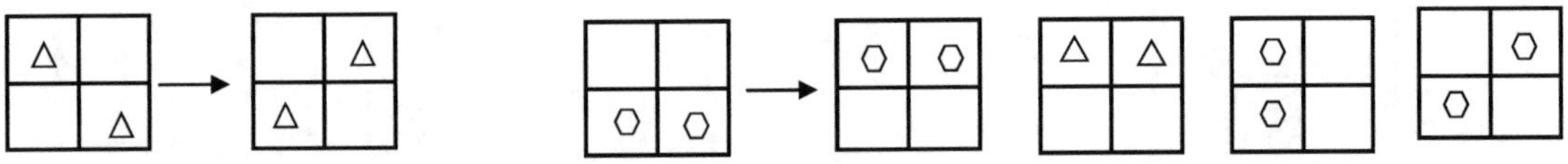

The triangles are in opposite corners of the grid of four squares. The triangles have been moved to the empty squares. The answer you should get is A.

How does it change? A B C D

1
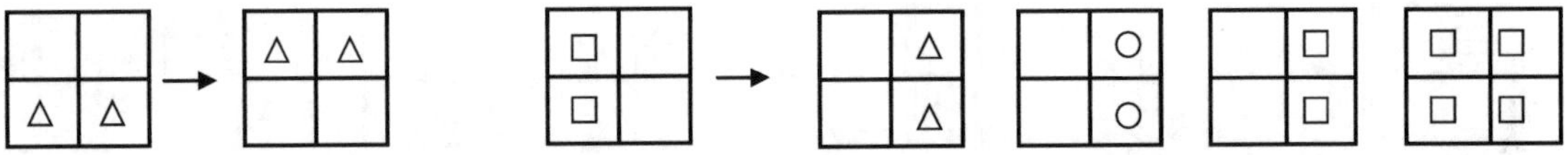

2
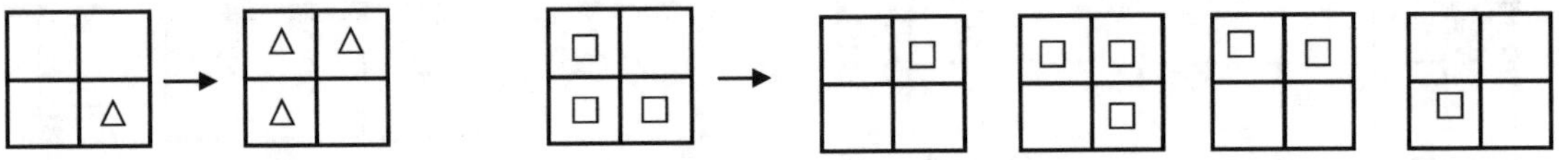

3
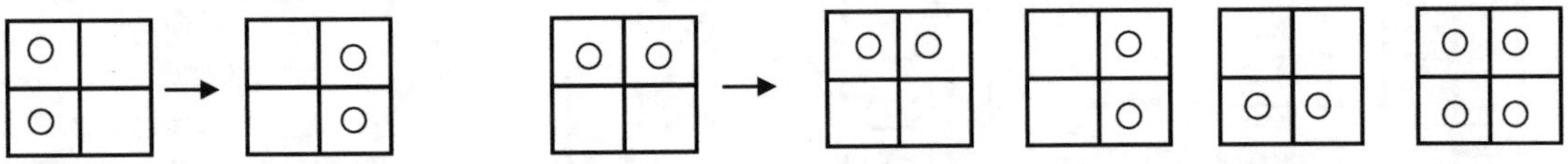

4
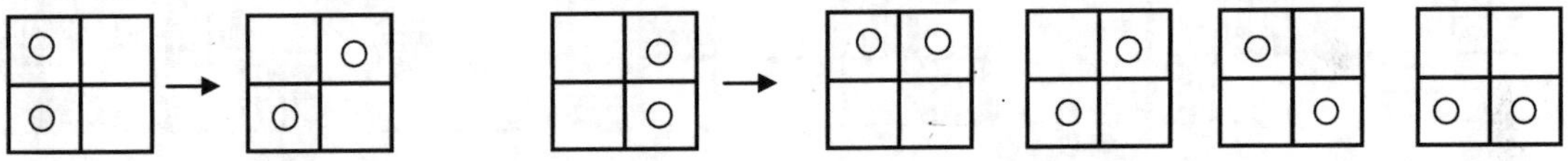

5
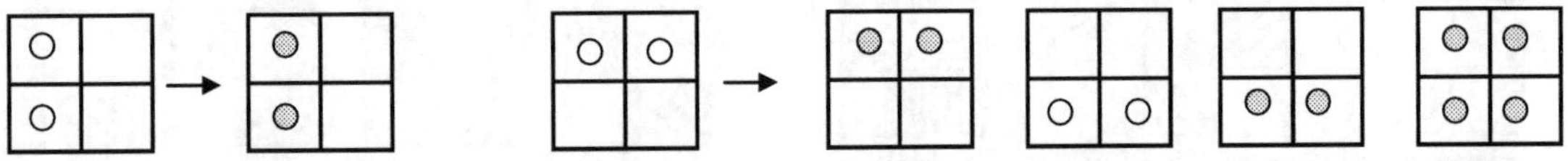

6
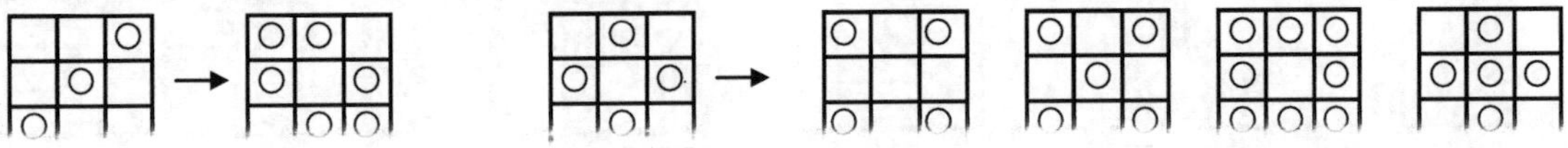

7
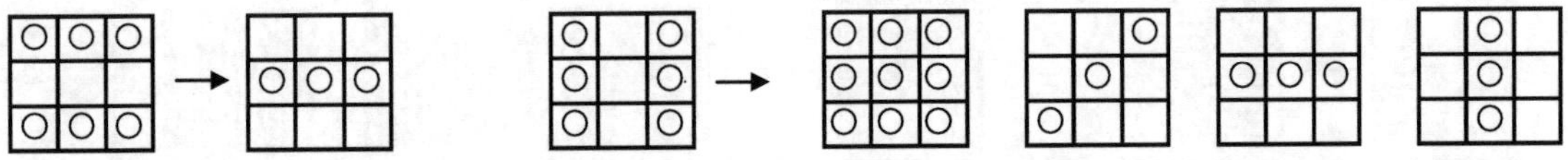

8
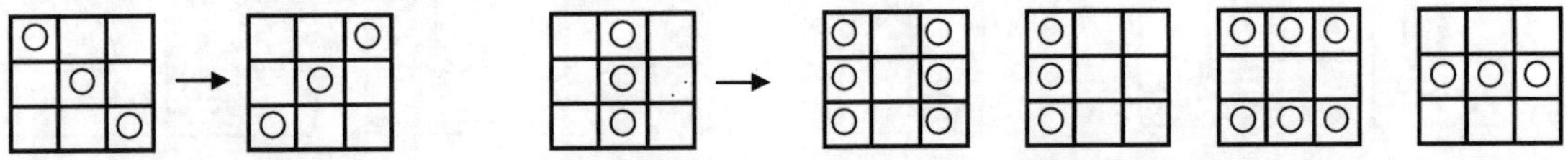

9
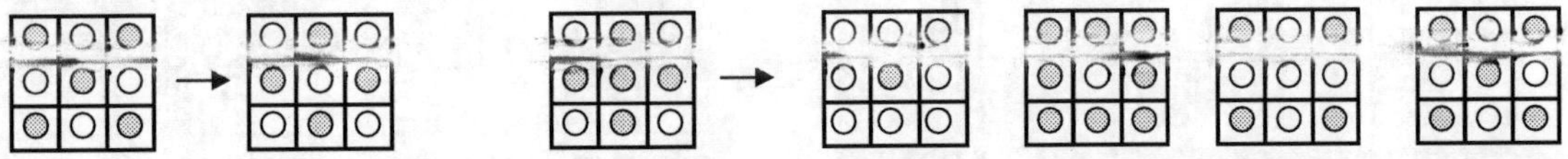

10
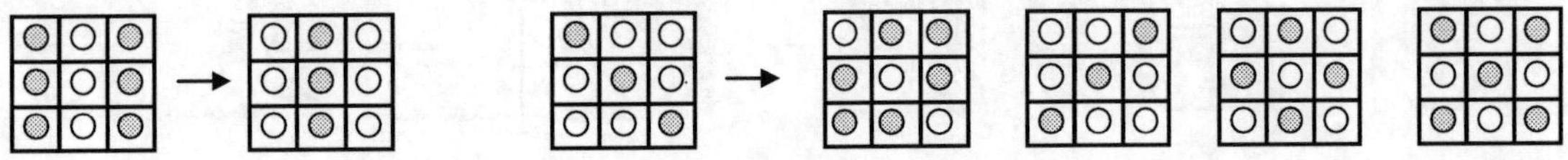

Mixed Non Verbal Worksheet 7.

There are two pattern in a grid where the first has been changed into the second. Look at the new grid and find the grid labelled A,B,C or D which would best fit if the shape had been changed in a similar way. Remember to use the same kind of change and then choose which one fits best.

Example: How does it change?

A B C D

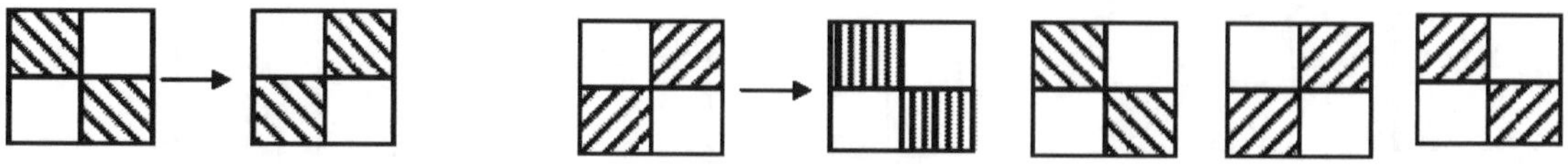

The shading are in opposite corners of the grid of four squares. The shadings have been moved to the empty squares and the direction of the shading lines travel in the same direction. The answer you should get is D.

How does it change?

A B C D

1 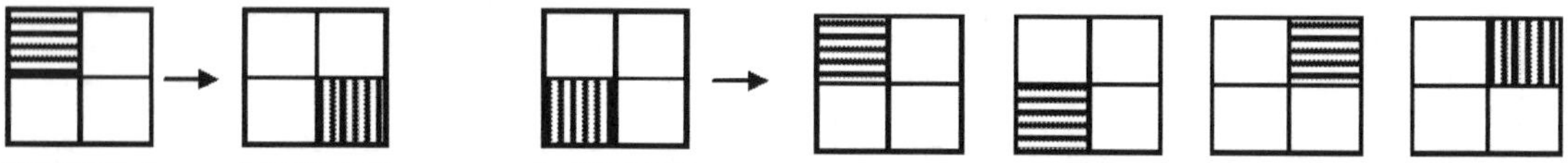

2 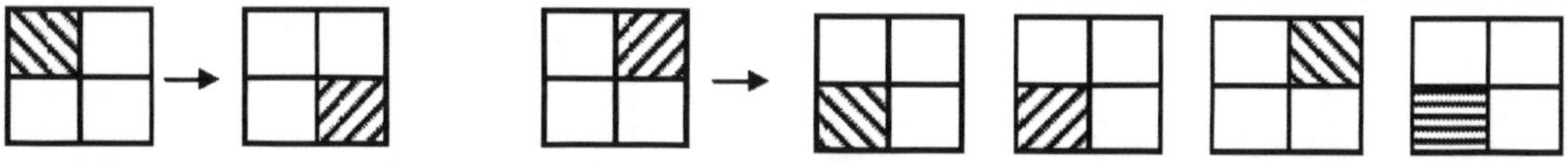

3 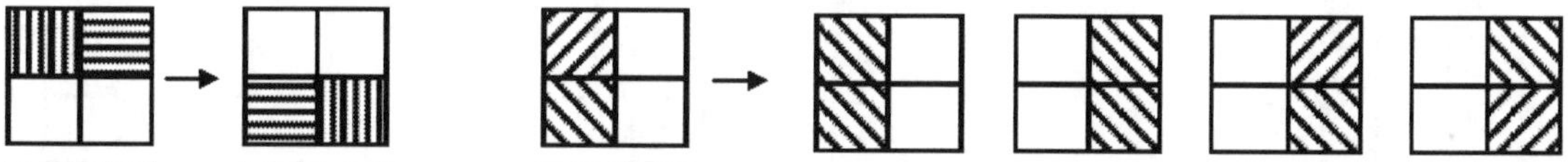

4 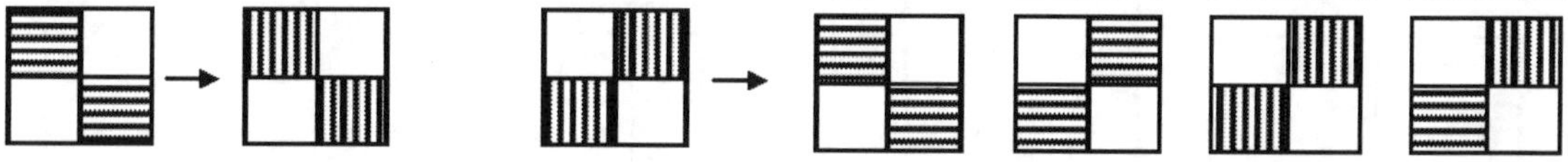

5 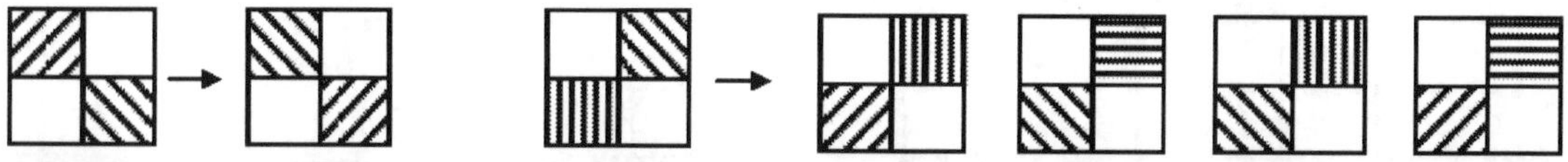

6 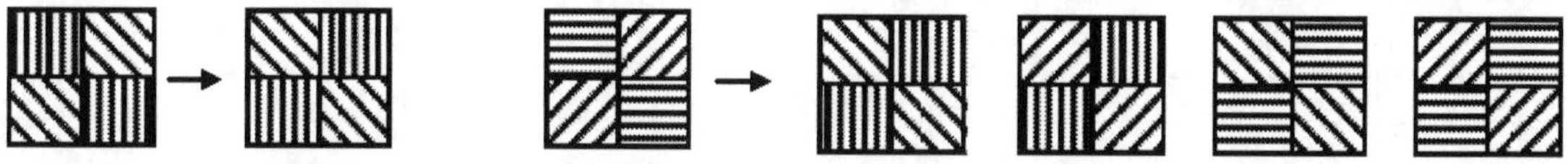

7 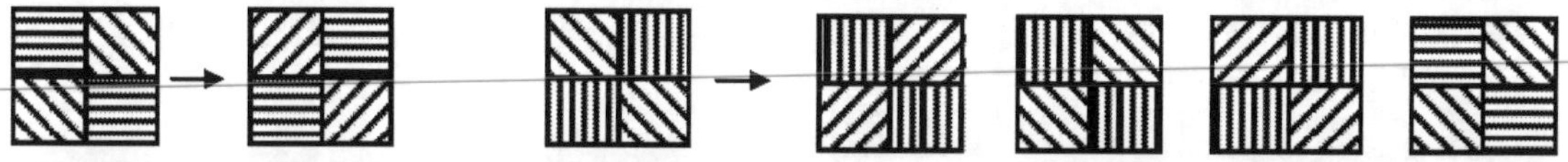

8 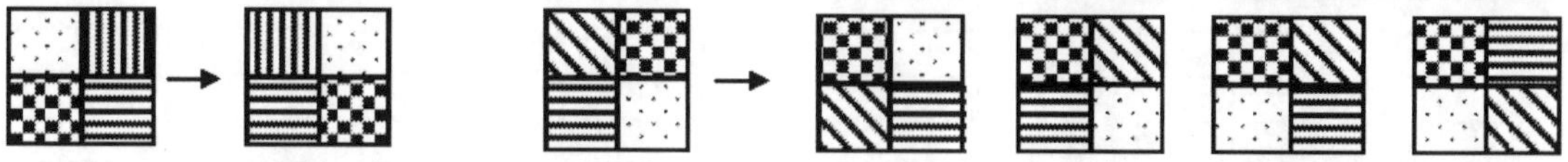

9 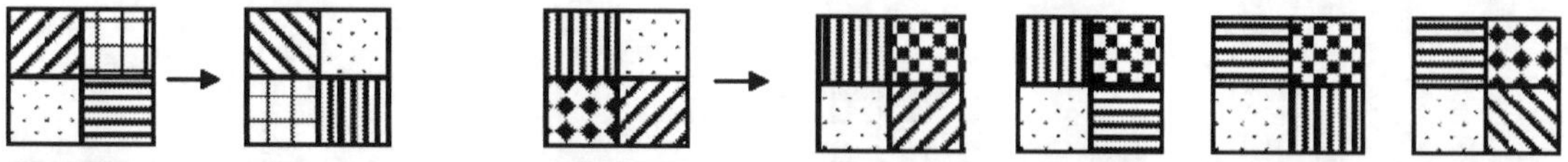

10 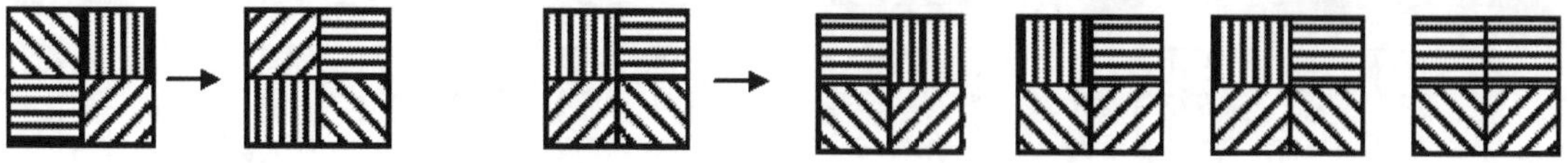

Mixed Non Verbal Worksheet 8.

There are two shapes where the first has been changed into the second. Look at the other shape and find the shape labelled A,B,C or D which would best fit if the shape had been changed in a similar way. You need to decide how they are changed in the example and make the same kind of change and then choose which one fits best.

Example: How does it change?

A B C D

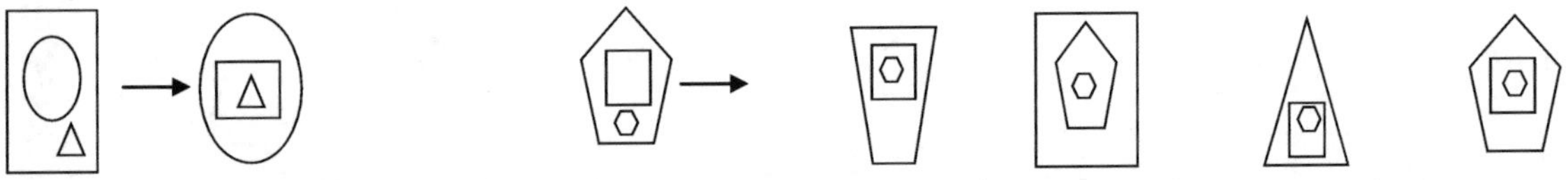

The outside shape which is a rectangle has become the middle. The larger of the two shapes inside the rectangle is an oval has become enlarged to the outside. The smallest shape is a triangle and stays the same size. Also the shapes are all inside each other. Follows the same rule for the other diagram and the answer you should get is B.

How does it change?

A B C D

1 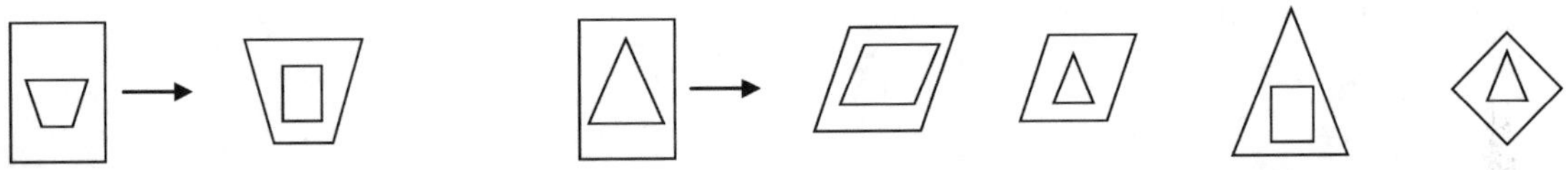

2 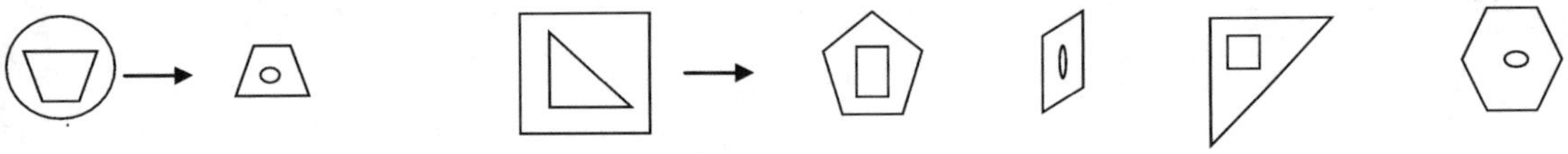

3 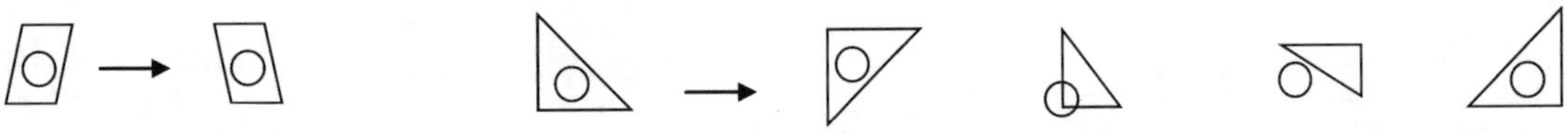

4 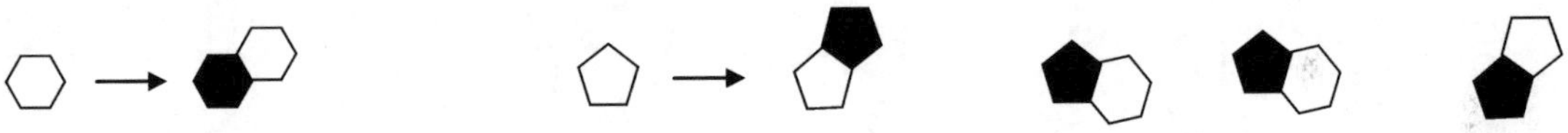

5 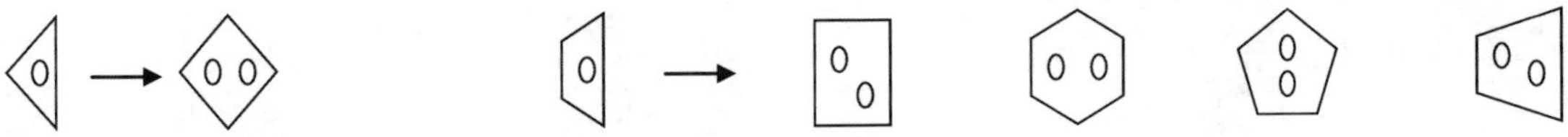

6 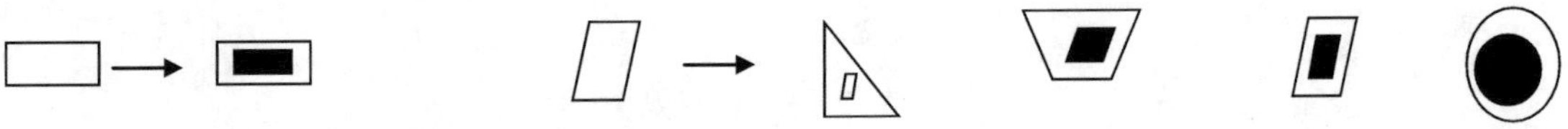

7 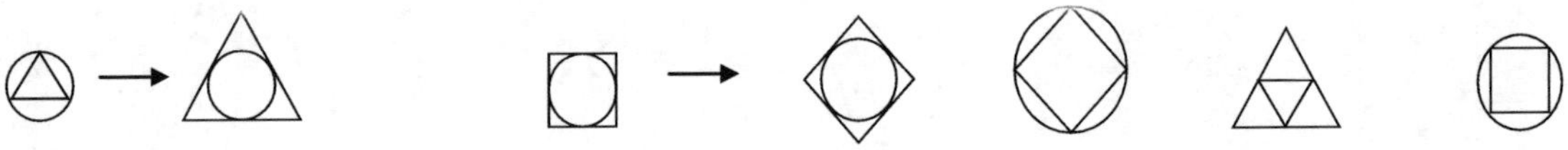

8 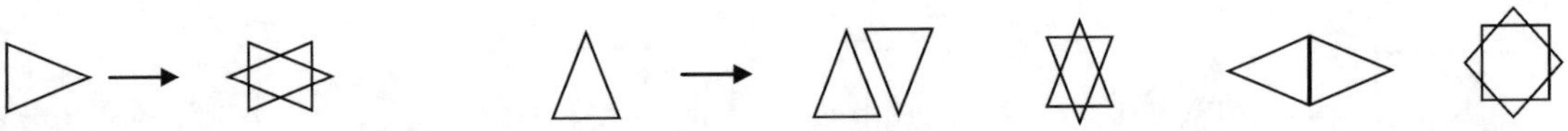

9 

10 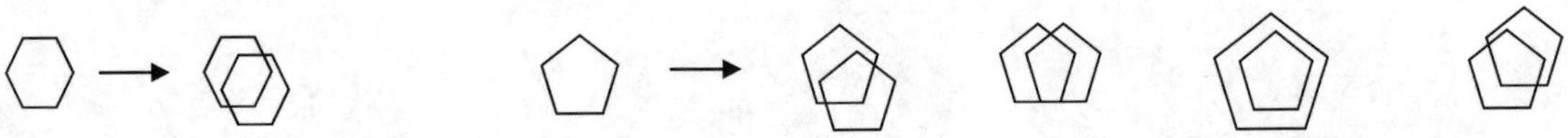

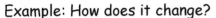

Mixed Non Verbal Worksheet 9.

There are two groups of shapes where the first has been changed into the second. Look at the new group of shapes and find the shapes labelled A,B,C or D which would best fit if the new group of shapes had been changed in a similar way.

Example: How does it change? A B C D

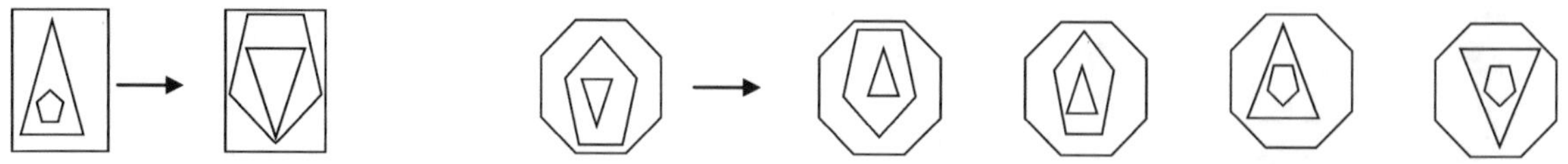

The outside shape is a rectangle and stays on the outside. The inside shape is a triangle which is rotated through half a turn and becomes the middle shape. The inside shape is a pentagon and is enlarged and rotated through half a turn and then fits outside the triangle but inside the rectangle, becoming the shape in the middle. The answer you should get is C.

How does it change? A B C D

1 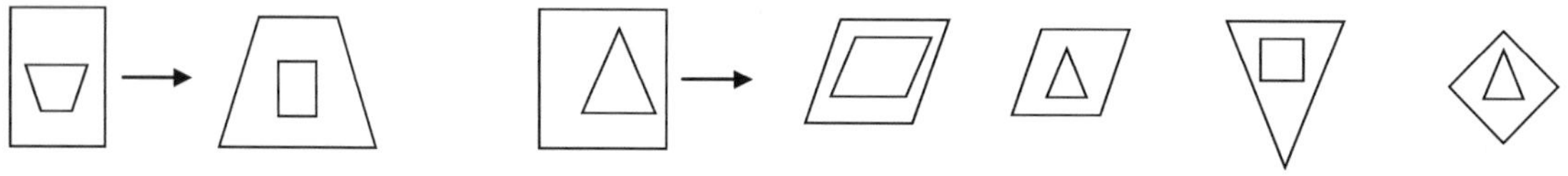

2 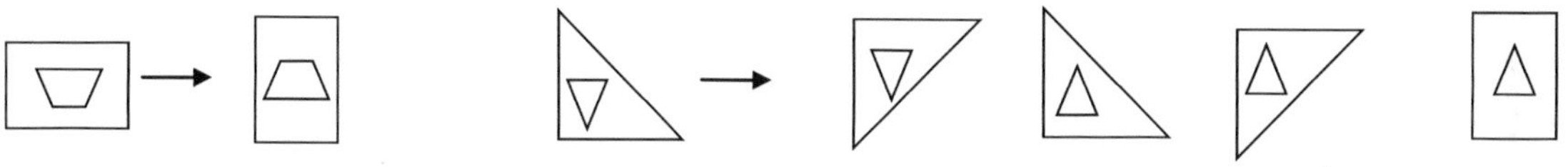

3 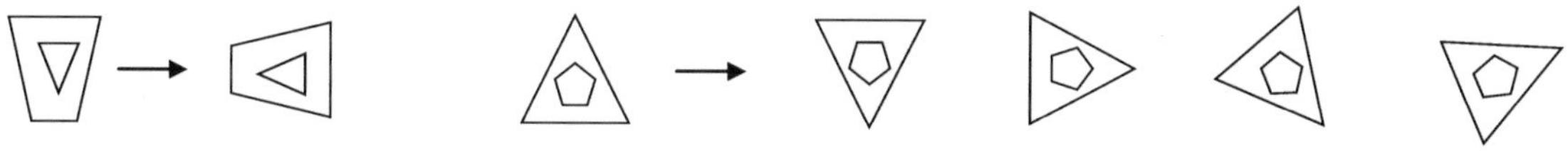

4 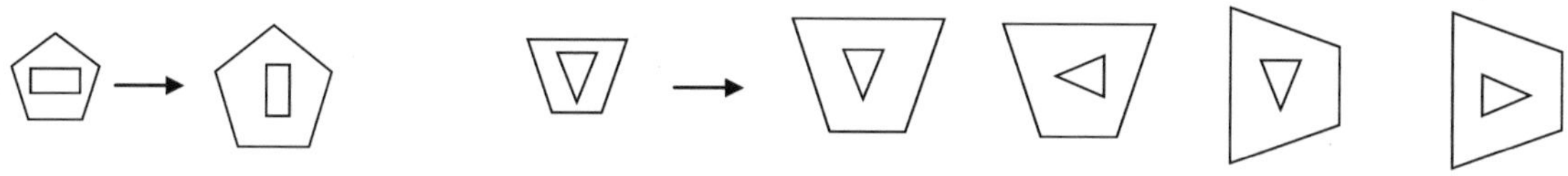

5 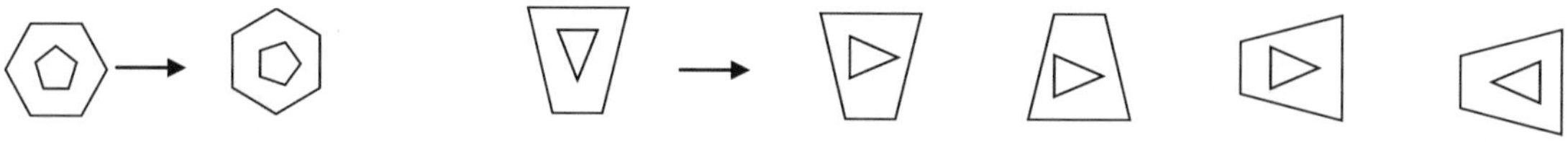

6 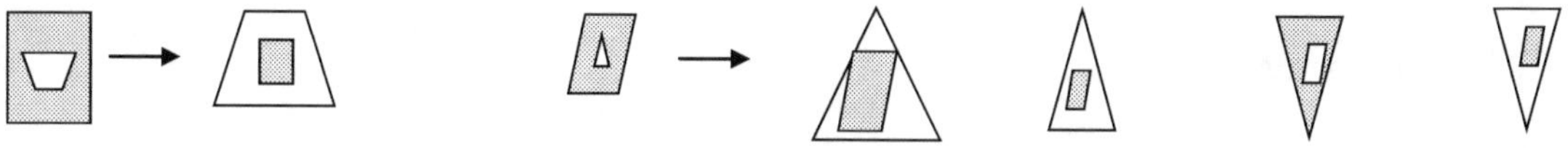

7 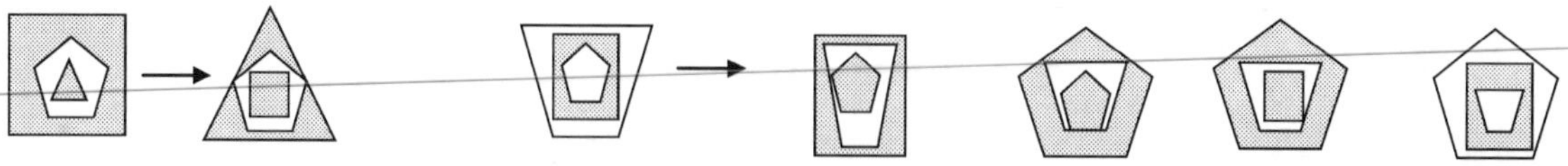

8 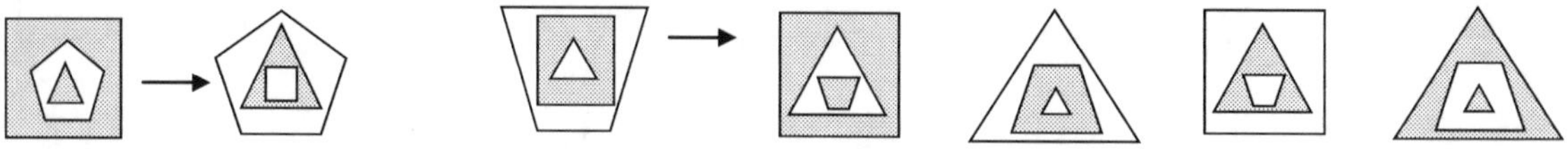

9 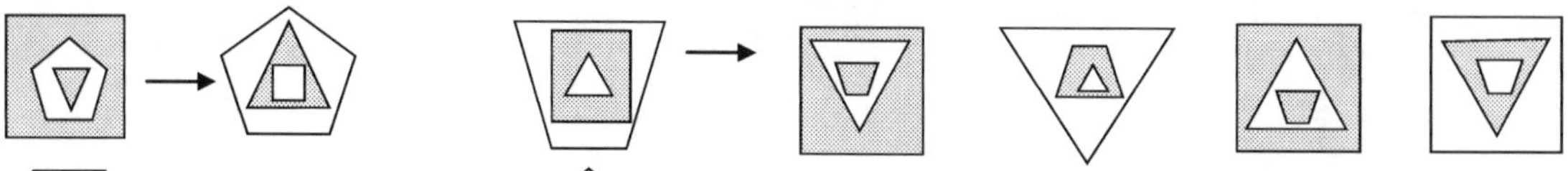

10 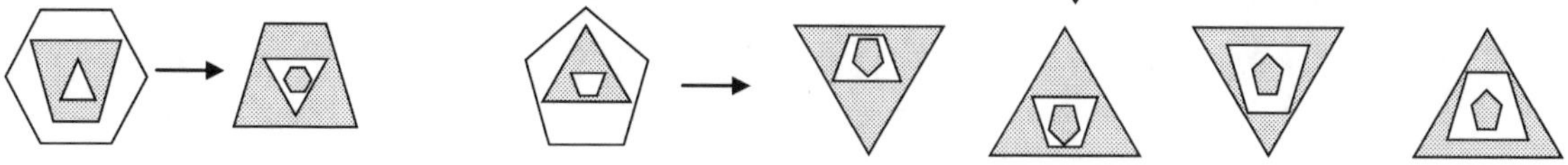

Mixed Non Verbal Worksheet 10.

There are two patterns of four shapes where the first set of four has been changed into the second set of four. Look at the new set of four shapes and find the correct set labelled A,B,C or D which would best fit if the shapes had been changed in a similar way. Remember to use the same kind of change and then choose which one fits best.

Example: How does it change? A B C D

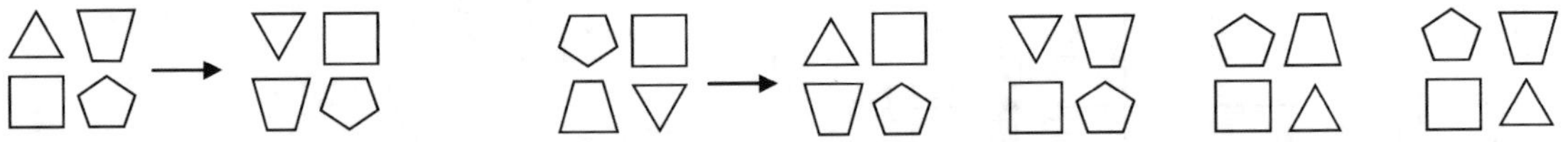

The triangle positioned in the top left has been rotated or turned upside down. The pentagon at the bottom right of the shapes has also been turned upside down. The trapezium and the rectangle have changed places. If you do the same to the new set of shapes the answer you should get is C.

How does it change? A B C D

1

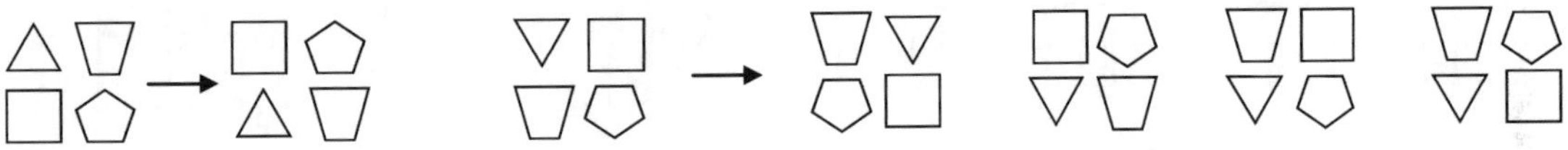

2

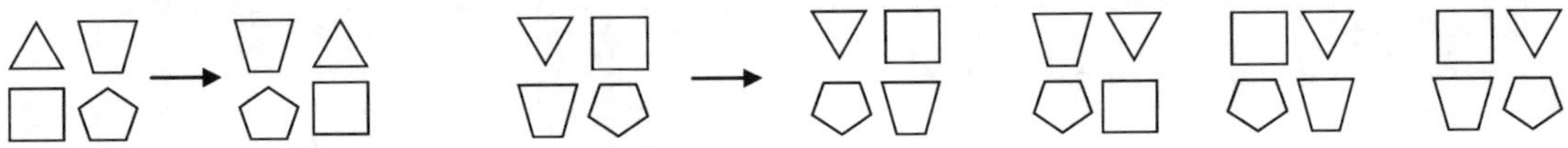

3

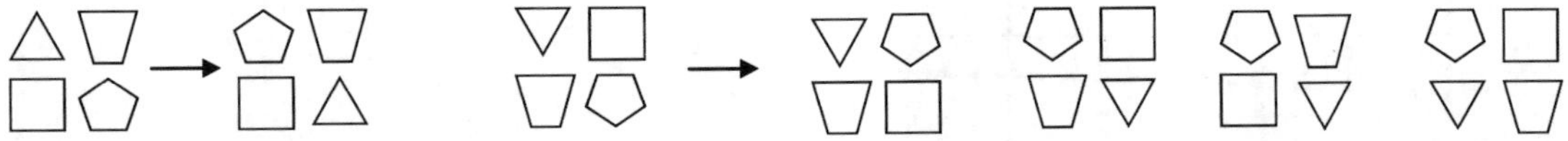

4

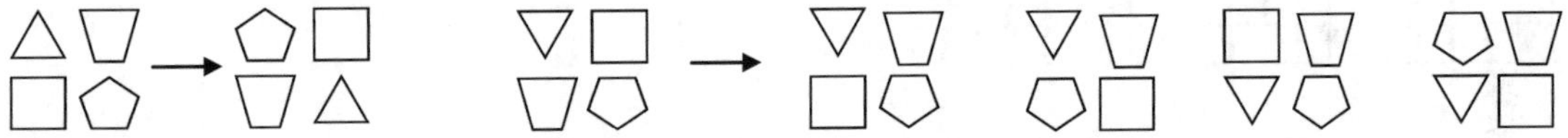

5

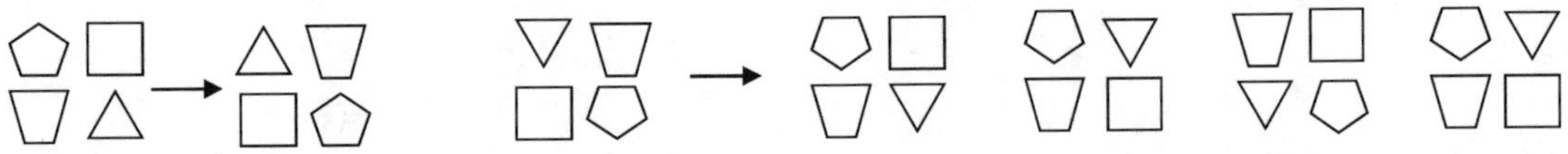

6

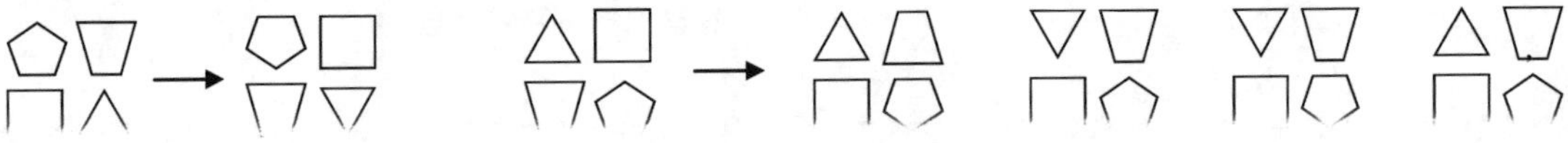

7

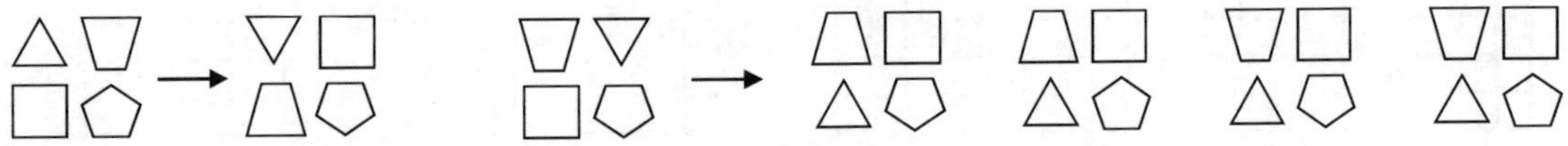

8

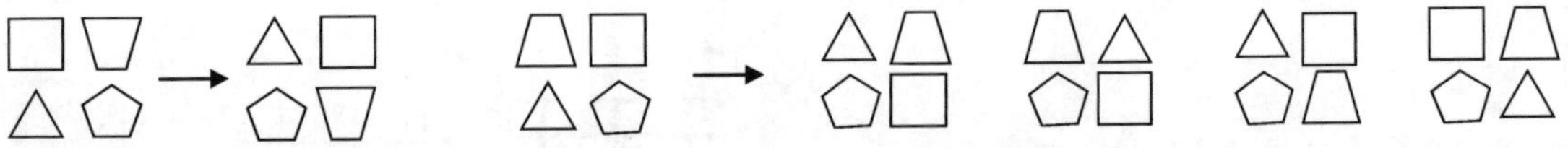

9

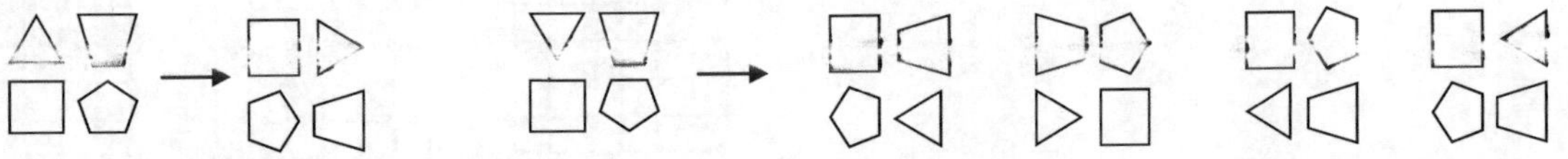

10

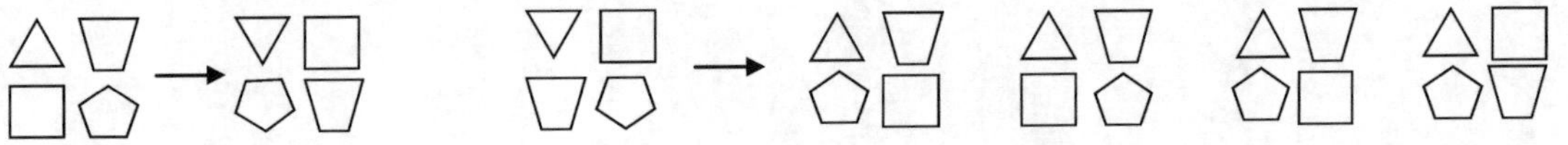

Answers to Activities, Practices and Worksheets.

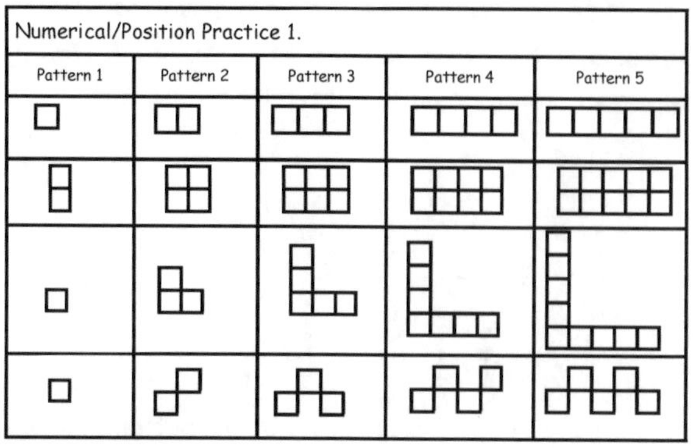

Numerical/Position Practice 2.

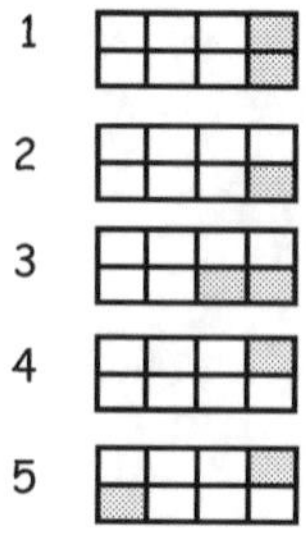

Reflection Practice 1 .

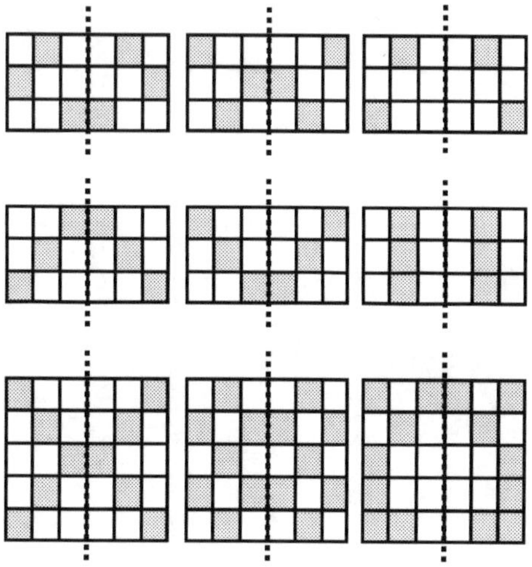

Reflection Practice 2.

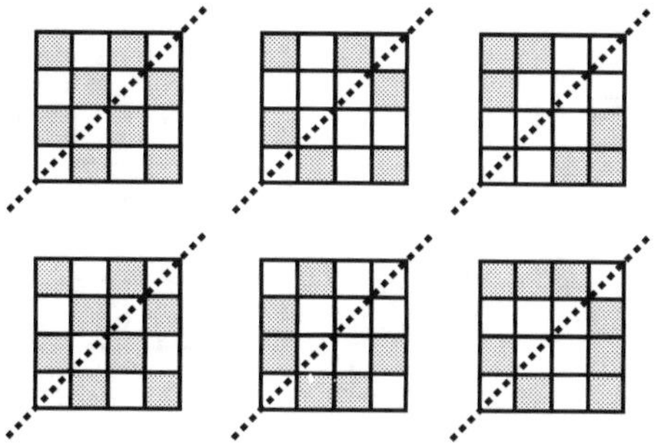

Reflection Practice 3.

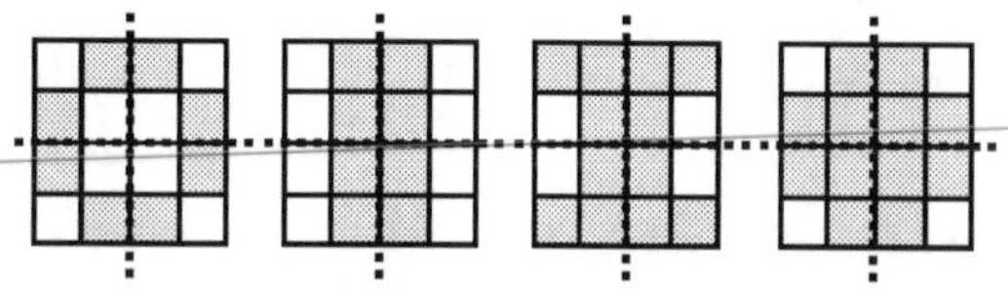

Reflection Practice 4.

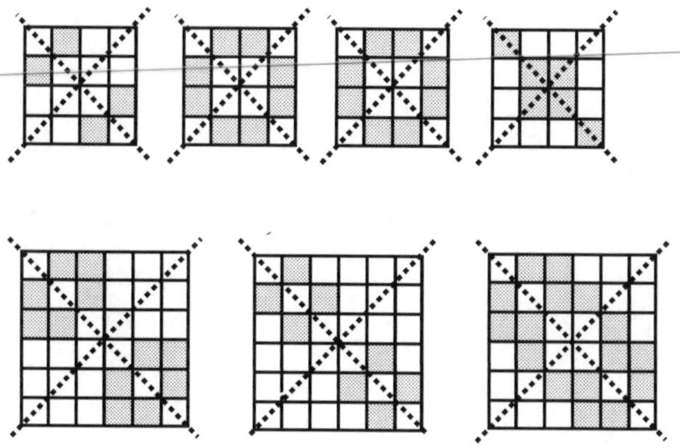

Rotation Practice 1

Here are the letters of the English alphabet that look the same when turned upside down.

Z H I N O S X Z

Rotation Practice 2

Here are the letters of the Greek alphabet that look the same when turned upside down.

Z H Θ I N Ξ O Φ X

Rotation Practice 3.

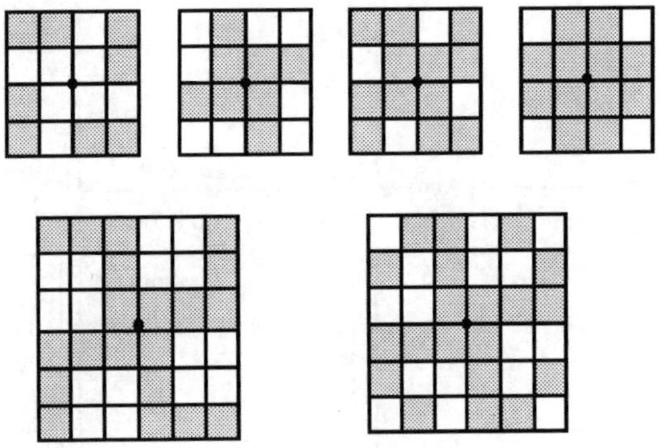

Rotational Practice 4.

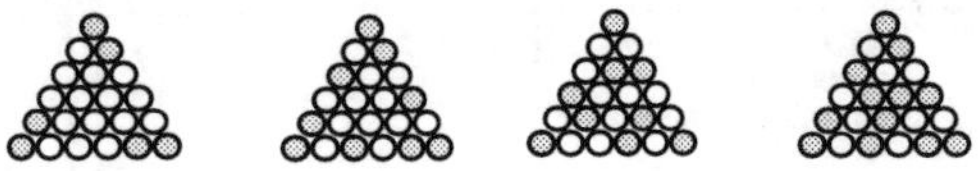

Similarity Practice 1.

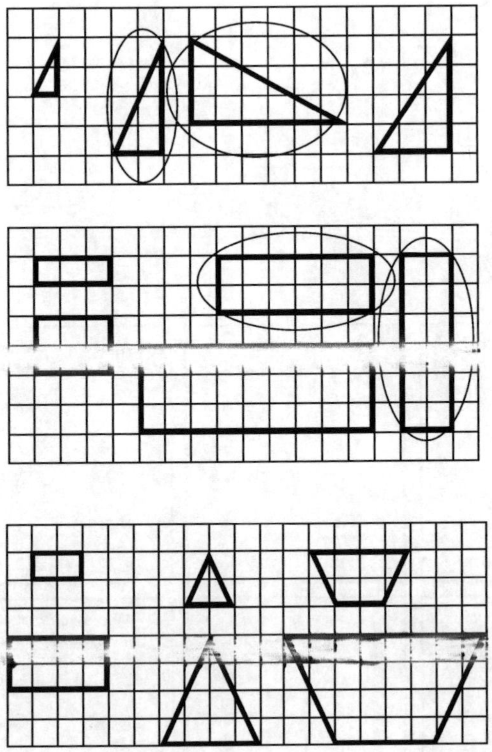

Similarity Practice 2.

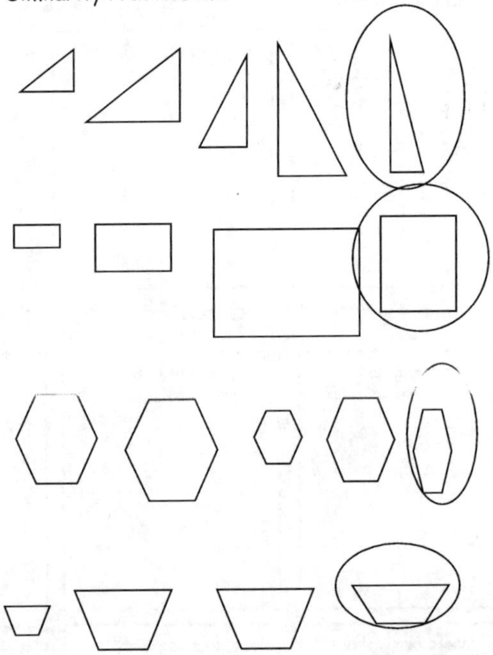

Section 1. Numerical/position patterns. Answers to worksheets.

Numerical Worksheet 1	
1 B	6 B
2 D	7 D
3 A	8 A
4 D	9 C
5 C	10 C

Numerical Worksheet 2	
1 B	6 D
2 C	7 A
3 C	8 B
4 B	9 D
5 A	10 C

Section 2. Reflection. Answers to worksheets.

Reflection Worksheet 1	
1 B	6 C
2 C	7 D
3 D	8 C
4 B	9 B
5 D	10 C

Reflection Worksheet 2	
1 C	6 C
2 D	7 A
3 E	8 E
4 C	9 B
5 D	10 A

Reflection Worksheet 3	
1 A	6 D
2 B	7 B
3 B	8 A
4 C	9 E
5 E	10 C

Reflection Worksheet 4	
1 B	6 B
2 D	7 A
3 C	8 D
4 D	9 B
5 B	10 D

Reflection Worksheet 5	
1 D	6 C
2 C	7 B
3 B	8 B
4 A	9 D
5 D	10 D

Section 3. Rotation. Answers to worksheets.

Rotation Worksheet 1	
1 A	6 D
2 D	7 B
3 B	8 A
4 A	9 A
5 C	10 B

Rotation Worksheet 2	
1 C	6 D
2 D	7 D
3 C	8 B
4 D	9 C
5 A	10 E

Rotation Worksheet 3	
1 B	6 E
2 C	7 C
3 E	8 C
4 A	9 B
5 C	10 B

Rotation Worksheet 4	
1 C	6 C
2 D	7 D
3 D	8 C
4 C	9 D
5 D	10 C

Section 4. Congruence and similarity. Worksheets.

Congruence and similarity Worksheet 1	
1 C	6 D
2 E	7 D
3 D	8 E
4 B	9 D
5 D	10 E

Congruence and similarity Worksheet 2	
1 C	6 A
2 D	7 B
3 B	8 C
4 A	9 C
5 C	10 D

Section 5. Non verbal Worksheets.

Mixed non verbal Worksheet 1	
1 C	6 A
2 D	7 D
3 B	8 C
4 B	9 D
5 A	10 A

Mixed non verbal Worksheet 2	
1 A	6 A
2 A	7 B
3 C	8 C
4 C	9 D
5 D	10 A

Mixed non verbal Worksheet 3	
1 C	6 C
2 B	7 A
3 C	8 A
4 A	9 A
5 D	10 B

Mixed non verbal Worksheet 4	
1 A	6 D
2 B	7 B
3 D	8 C
4 C	9 D
5 B	10 B

Mixed non verbal Worksheet 5	
1 C	6 D
2 A	7 C
3 D	8 C
4 D	9 A
5 C	10 B

Mixed non verbal Worksheet 6	
1 C	6 B
2 A	7 D
3 C	8 D
4 D	9 C
5 A	10 A

Mixed non verbal Worksheet 7	
1 C	6 D
2 A	7 B
3 D	8 C
4 B	9 D
5 C	10 A

Mixed non verbal Worksheet 8	
1 C	6 C
2 C	7 D
3 D	8 B
4 D	9 D
5 B	10 A

Mixed non verbal Worksheet 9	
1 C	6 D
2 C	7 D
3 B	8 A
4 B	9 A
5 D	10 A

Mixed non verbal Worksheet 10	
1 D	6 C
2 C	7 B
3 B	8 A
4 A	9 D
5 A	10 A

ABOUT THE AUTHOR

Nigel Robinson has an honours degree in mathematics and physics and has a post graduate teaching diploma in Education.

He has been teaching mathematics in secondary and high schools for 25 years and has coached students in the enhancement of reasoning and thinking skills and techniques to enable students to pass entry examinations for selective schools and colleges. The author has also coached students who wished to become members of Mensa. The material, worksheets, and activities used by the author is compiled by him and so meets the needs of the individual students.

Printed in the United States
By Bookmasters